The Hitchhiker's Guide To Clinical Chemistry

Dr. I. Wilkinson, Ph.D., D. Clin. Chem., FCACB

Department of Clinical Biochemistry
Sunnybrook Health Science Center
2075 Bayview Avenue
Toronto, Ontario
Canada, M4N 3M5
(416) 460-4991 • FAX: (416) 480-6120

This guide is supposed to be just that: a guide.
It is not an encyclopedia, nor is it a bible.
It was written as a quick guide to common tests and ideas—
a toolbox for students taking clinical
biochemistry courses for fun and profit.
Reference ranges, other values and procedures described
in this book may vary between different institutions.
Whilst care has been taken to eliminate
errors the author does *not* warrant that
the information contained herein is in every respect
accurate or complete and is not responsible
for any errors, or omissions, or for the results
obtained from use of such information.
This book is not to be taken internally.
Keep out of reach of small children and bored students.

Ian Wilkinson, July 1992

For Mam and Dad and Sandra

Acid-Base

The body seeks to maintain a pH of 7.35-7.45 units [_ pH]. Life is inherently dangerous. We are all teetering on the edge of acidosis (our pH is constantly in danger of dropping) due to the production H^+ ions from metabolism. Henderson and Hasselbalch came up with the following equation:

$$pH = pK + \log \quad \frac{[HCO_3]}{0.03 \times pCO_2}$$

...which is just 'Mathspeak' for saying that in general:

$$pH \text{ is proportional to} \quad \frac{[HCO_3]}{pCO_2} = \frac{Kidney}{Lung}$$

To end up with a pH = 7.35-7.45, it turns out that the body has to keep the ratio of $[HCO_3^-]$ to pCO_2 at 20:1. Notice that the actual amounts of bicarbonate and dioxide do not matter (within reason!) as long as the ratio of the two substances is close to 20:1, then the pH will be about right for life to go on.

The kidney controls pH by excreting or retaining hydrogen ions. The lungs maintain pH by excreting or retaining CO_2. Only the kidney can actually excrete H^+ ions, the lungs can only move the equation to left or right and therefore help to keep the ratio of $[HCO_3^-]$ to pCO_2 as close to 20:1 as possible.

$$HCO_3^- + H^+ \rightarrow H_2CO_3 \rightarrow CO_2 + H_2O$$

Acid-base diseases arise when the body no longer maintains the ratio at 20:1. Acid-base diseases are either due to an increase in pH = 'alkalosis' or a decrease in pH = 'acidosis'. Either alkalosis or acidosis can be caused by changing the denominator (pCO_2) or the numerator (HCO_3^-), i.e., because of respiratory problems or renal (metabolic) problems. Thus, you could have four types of acid-base problems: respiratory and renal (usually called 'metabolic' or 'non-respiratory') acidoses and respiratory and non-respiratory alkalosis; depending upon what was the primary cause. The

1

body always tries to compensate for the primary change, e.g, if the bicarbonate starts to fall for some reason, then the body will try to decrease the pCO_2 (by making you hyperventilate) thereby attempting to get the ratio back toward 20:1 (pH = 7.35-7.45). The lungs can compensate almost immediately. The kidneys require much longer to compensate (many hours to days).

ACTH

[☛ Adrenocorticotropic Hormone]

Addison's Disease

Primary adrenocortical insufficiency, i.e., the adrenal cortex is unable to produce sufficient aldosterone and cortisol [☛ Aldosterone, Cortisol]. Autoimmune mechanisms and tuberculosis are known causes of Addison's disease.

Symptoms and signs reflect the loss of the above hormones. hypotension, hyponatremia, hyperkalemia (aldosterone deficiency) and skin pigmentation, defective carbohydrate metabolism (cortisol)

Adrenocorticotropic Hormone (ACTH)

Reference range (at 08:00 a.m.): 4.-31 pmol/L (20-140 pg/mL)

ACTH is produced by the anterior pituitary in response to stimulation by corticotrophin releasing factor (CRF) (released from the hypothalamus). ACTH acts on the adrenal glands causing an increases in cortisol concentration in the blood [☛ Cortisol].

As the concentration of cortisol increases there is a negative feedback effect by cortisol on the release of ACTH. CRF, ACTH and cortisol all exhibit diurnal variation. Random samples must therefore be collected at a consistent time of day, preferably when levels are expected to be at or close to their peak (8:00 a.m.) By 24:00 hrs levels will be approximately one-half of those seen at 8:00 a.m. ACTH is useful in the differential diagnosis of Cushing's Syndrome [☛ Cushing's Syndrome].

2

Alanine aminotransferase (ALT)

Reference range: 0-30 IU/L {0-30 IU/L}

Also known as Serum Glutamic Pyruvate Transaminase (SGPT) (at least in extinct text books and parts of the U.S.). ALT tends to increase whenever cells of certain organs (liver and heart especially) are killed or badly damaged. It is also increased in about 60% of patients that receive heparin therapy. ALT (as its name implies) transfers amine groups:

alpha-ketoglutarate + alanine —^{ALT}—> glutamate + pyruvate

Albumin

Reference range: 35-50 g/L {3.5-5.0 g/dL}

One of the major proteins in the blood. Albumin is a major components of oncotic pressure, i.e., the concentration of albumin in the blood vessels tends to 'attract' water from the interstitial spaces of the tissues into the blood vessels (vascular space). In disease (chronic liver disease {albumin is synthesised in the liver} malnutrition, malabsorption, nephrotic syndrome, severe burns) where albumin concentration falls considerably, this oncotic pressure is reduced and water tends to move into the interstitial space and remain there, resulting in edema.

'Albumin' is derived from the Latin word 'albus' meaning white. Albumin is a major constituent of the white of an egg (albumen). Albumin was probably spelled, 'albumine' originally: meaning a substance derived from albumen.

Aldosterone

Reference range: 100-500 pmol/L {(3.6-18 ng/dL} recumbent: depends upon posture, gender, sodium in diet etc).

Hormone produced in the adrenal cortex in response to increase in angiotensin [☛ Renin-Angiotensin System]. Function: increases sodium reabsorption and potassium or H^+ secretion in the distal tubules of the kidney. Increased sodium reabsorption results in increased water reabsorption and therefore to hypertension. Increased aldosterone causes a fall in renin due to feedback. Not to be confused with the well known Sicilian mobster Al D'O. Steroni.

Alkaline Phosphatase (ALP)

Reference range: 30-90 IU/L {30-90 IU/L}

Total ALP may be increased as part of normal physiological development and in disease states involving liver, intestine, bone or placenta. Total ALP is composed of ALP isoenzymes which are tissue specific for each of these organs/systems [☞ Isoenzymes, Electrophoresis]. In liver cholestasis, for example, ALP_{liver} isoenzyme is increased by induction (not due cell necrosis or damage as is the case for AST and ALT [☞ AST, ALT]).

Pubescent teenagers show increased levels of ALP_{bone} during the period of growth. Pregnant women (last trimester) show elevated $ALP_{placenta}$ and post-menopause (ALP_{bone}). Thus, reference ranges must be adjusted for these classes of patient.

ALP

[☞ Alkaline Phosphatase]

ALT

[☞ Alanine aminotransferase]

Amylase

Reference range: 40-160 U/L {40-160 IU/L}

Amylase is an enzyme that breaks down carbohydrates into simpler units. The pancreas releases amylase into the small intestine in response to the enzyme secretin [☞ Secretin]. Secretin (produced by specialized cells in the walls of the small intestine) is stimulated by the presence of food in the duodenum.

Amylase is found in several organs of the body (pancreas, salivary glands, testicles) [☞ Lipase]. In children increased amylase is often a result of mumps or other causes of inflammation of the salivary glands.

4

In adults mumps is far less likely. Increase in amylase due to pancreatitis (stones, infection, alcohol, tumor, perforated duodenal ulcer) may lead to increased serum amylase levels within 2-6 hours, peaking at 12-30 hours and remains elevated for 2-4 days.

The serum amylase is eventually excreted by the kidneys, resulting in an increase in urine amylase. In acute pancreatitis the renal glomerular filtration of amylase actually increases! This means that amylase is often elevated in the urine several days after the serum amylase levels are 'normal' (within serum reference range).

This can be useful in diagnosing patients who do not present themselves to a physician, immediately following an episode of acute pancreatitis. By measuring how much amylase is cleared from the serum (amylase clearance) it is possible to help in differential diagnosis of acute pancreatitis.

'Amylase' is derived from 'amylos', the Greek word for, 'starch' (a carbohydrate). The '-ase' ending is 'Chemspeak'; for 'enzyme'.

Analogy

Reference range: 'its like something or other'

Not to be confused with an allegory or an allergy. The former uses a story to teach a moral or give a message. The latter is a plot by people with shares in paper tissues. Many people are severely allergic to ragweed and income tax. Analogy on the other hand is little more than a devious educational tactic: using something familiar to describe and explain something unfamiliar.

Angiotensin

[see Renin-Angiotensin System]

Anion Gap

Reference range: 12-18 mmol/L* (12-18 mEq/L)

Anion Gap = $[Na^+] - ([Cl^-]+[HCO_3^-])$

By substituting the reference range values for sodium (135), chloride (95) and bicarbonate (22), you can see that the 'reference value' for a 'normal' anion gap is about 12-18 mmol/L. Do not confuse with 'osmolar gap' [☛ Osmolar Gap].

*Note that some labs include potassium (3.5) giving a different 'normal' value.

What use is the anion gap? In metabolic acidosis, if a <u>high</u> anion gap is observed, you can be almost certain that the only one of four clinical conditions is responsible: renal failure, ketoacidosis, drugs/toxins or lactic acidosis.

A <u>low</u> anion gap is seen in multiple myeloma and hyponatremia etc.

'Chemspeak' can be somewhat confusing when it comes to ions: anions are *negatively* charged ions. Cations are *positively* charged ions. But, an <u>anode</u> is a positive electrode and a <u>cathode</u> is a negative electrode. Anions are ions that are attracted to anodes. Since opposites attract (at least in Scienceville they do!), negatively charged ions are attracted to the positively charged anode and are therefore called anions. The converse applies to cations. The anion gap is not really a gap at all. It is an invention. A tool that aids in diagnosis. People are neutral when it comes to electric charge. If they were not, you would spend a lot of time sticking to walls and ceilings, getting covered in dust, paper, plastic, hairs etc; and dating would be highly dangerous. The missing anions (negatively charged) are present in but not measured. Talking about an increase in the so-called gap is really the same as saying that there is an unexplained increase in positively charged ions (cations), e.g., H^+ ions; therefore there are more pluses than minuses and the gap increases.

The word 'ion' was invented by Michael Faraday (1791-1867), (except that his contemporary, William Whewell, may also have invented it!). 'Ion' is derived from the Greek word, 'ienai' meaning 'to go'. 'Ana' is Greek for 'up': hence 'anion' ' means 'to go up'. The word 'cation', as you might have already guessed, means the opposite: 'to go down'. (From the Greek, 'kata', meaning,'down'). Anions, cations, cathodes and anodes are all examples of 'Chemspeak' invented words and terms referred to as neologisms [☛ Iatrish].

6

Anti-Diuretic Hormone (ADH) aka 'Vasopressin'

Reference range: 0-10 ng/L {0-10 pg/mL}

ADH is released from the posterior pituitary (although it is actually synthesized in the hypothalamus and then stored in pituitary) and acts on the collecting ducts of the kidney: it facilitates the retention of water - hence the name anti-diuretic (opposing a diuresis or 'flow'). ADH release is triggered (1) when osmoreceptors (in the hypothalamus) detect an increase (changes of about 2%) in osmolality and (2) when baroreceptors (in the right atrium {heart} and the carotid sinus) detect a fall (changes of about 10%) in intravascular volume (pressure also falls). Note that the body gives volume control higher priority than osmolality control if it must choose between the two. (3) ADH is also increased during stress (surgery, trauma), exercise, erect posture and at night-time. [☞ Syndrome of Inappropriate secretion of Anti-Diuretic Hormone (SIADH)].

ADH and Growth Hormone are two 'Chemspeak' terms that actually tell you something of what they do. These must have slipped through the *International Committee for Giving Things Totally Confusing and/or Upronouncable Names That Have Little or Nothing to Do With Anything.* (for examples of better work by this committee [☞ Creatine and Creatinine or just about anyone's syndrome].

Aspartate aminotransferase (AST)

Reference range: 0-30 IU/L {0-30 IU/L}

Also known as Serum Glutamic Oxalacetic Transaminase (SGOT) in older, extinct text books and in parts of the U.S. As its name implies AST transfers amino groups:

alpha-ketoglutarate + aspartate —^{AST}—> glutamate + oxalacetate.

AST is widely distributed in various tissues throughout the body. An increase in AST is a consequence of cell death or damage: it 'spills out' from the damaged or dead cells. AST is often increased in: liver diseases involving hepatocyte necrosis; in heart in acute myocardial infarction (MI), ('heart attack'); skeletal muscular diseases; red blood cells (severe hemo-

7

lytic anemia); malignancies etc. Thus an increase in AST is not by itself diagnostic of, e.g., liver disease or MI. If ALT is also increased one would suspect liver disease rather than heart [Alanine aminotransferase (ALT)].

AST

[☞ Aspartate aminotransferase]

Bicarbonate

Reference range: 22-30 mmol/L {22-30 mEq/L}

Carbon dioxide, CO_2 exists largely as HCO_3^- in solution. This reaction is catalyzed by carbonic dehydratase, in red blood cells and the lumen of the nephron. Bicarbonate is reabsorbed in the proximal tubule. The reabsorption is indirect, i.e., it is converted to CO_2 via carbonic dehydratase. The ratio of concentration of bicarbonate to carbon dioxide determines the pH of the blood (see Acid-Base). Bicarbonate is one of the major bases that buffers hydrogen ions in the body.

Actual CO_2 is the concentration of bicarbonate calculated by a blood-gas analyzer. [☞ Blood Gases].

Total CO_2 is the sum of HCO_3, CO_2(dissolved) and CO_2 bound to various proteins (hemoglobin etc).

'Bicarbonate' is derived from 'bi' meaning 'two' and 'carbo' Latin for 'charcoal'. The 'ate' bit at the end is 'Chemspeak' for 'O_3'.

Bilirubin

Reference range: 2-20 mol/L {0.1-1.2 mg/dL} (total bilirubin)

Total bilirubin = Bu + Bc
Bu = unconjugated bilirubin (aka: 'indirect' in older texts)
Bc = conjugated bilirubin (aka: 'direct' in older texts) = BMG + BDG
BMG = bilirubin monoglucuronide
BDG = bilirubin diglucuronide

Bilirubin is a breakdown product of the heme part of hemoglobin (and from other sources: ineffective erythropoiesis, myoglobin, cytochromes—all contain the heme group): senescent red blood cells —> heme + globin (globin is broken down and its amino acids re-used. The iron (Fe) atom at the center of the heme molecule is removed and recycled by the body (very environmentally friendly of it!). It is inefficient, in terms of energy required, for the body to breakdown the heme moiety, it is therefore modified —> bilirubin. This is <u>un</u>conjugated bilirubin (Bu) and is insoluble which is a pity, because blood is mainly water and the bilirubin has to travel down the intravascular highway of the blood vessels to reach the liver (not the kidney as you might expect!) and be excreted. The body gets around the insolubility problem by albumin and Bu getting together but non-covalently. Bu-albumin in association is soluble. This 'complex' travels to the liver. The albumin leaves the Bu and the latter is transported into the hepatocyte where it is chemically modified: it is conjugated in two steps to form firstly bilirubin monoglucuronide (BMG) and finally bilirubin diglucuronide (BDG). Conjugated bilirubin (Bc) consists mainly of BDG. Most of the Bc now passes into the bile duct and eventually into the intestine. Much of the bilirubin is metabolized by the local intestinal flora which firstly de-conjugate the Bc—>Bu (the losers!) then convert it to urobilinogen and then to urobilin which is excreted in the feces. It is the urobilins that give the feces that familiar color scheme known as U*taup*ia.

That is the basic story. There are one or two details to add: remember that the body constantly absorbs things that are present in the small intestine (ex-hamburgers, ex-quiche etc). It also absorbs some of the Bc and urobilinogen that are also in the intestine. Absorbed nutrients go to the liver and are utilized by the body, and the Bc and urobilinogen go with them, only to be once again excreted via the bile duct (I wonder if they get a feeling of déjà vu?). This whole process is known as the enterohepatic circulation. A small amount of the Bc and urobilinogen can find its way into the general circulation and eventually reaches the kidneys and is excreted in the urine. In health the amounts in the urine are very small. In disease, e.g., hepatitis, levels of Bc and urobilinogen in the urine may be increased.

Bc in the blood stream may also be associated with (NOT covalently bound to!) albumin but more loosely than Bc associates with albumin. Only Bc can appear in urine. Bu is too tightly associated with albumin and the Bu—albumin complex it too large to pass through the glomerulus.

Total bilirubin may increase in: liver disease (hepatitis), hemolytic anemia, congenital diseases (Crigler-Najaar syndrome, Dubin-Johnson Syndrome etc).

'Bilirubin' is derived from the Latin words, 'bili', meaning, 'bile' and, 'rubin' meaning, 'red'. This piece of 'Chemspeak' was to distinguish it from yellow and green analogs; and seems almost logical until you realize that bilirubin is not red in color but yellow! Ours not to reason why.

Blood Gases

Blood gases includes pH (not a gas!), pCO_2, pO_2 and bicarbonate (also not a gas!—yet more 'Chemspeak'!)

[☛ Acid-Base, pH, pCO_2, pO_2 and Bicarbonate].

Note that these tests are generally performed in <u>arterial</u> blood not on venous samples as are the majority of biochemical tests. Arterial samples require a physician due to the inherent dangers of tapping into an artery as opposed to a vein.

Breath Hydrogen Test for Bacterial Overgrowth

Reference range: ≥12 parts per million = positive for bacterial overgrowth.

People do not make hydrogen. Intestinal flora do. Hydrogen is the end product of bacterial metabolism. The hydrogen is given off by the bacteria present in the lumen of the large intestine. They are responsible for much of the pain, noise, aroma, humor and spontaneous human combustion associated with flatulence. Not all hydrogen passes out of the body through the anus. Some of it diffuses through the wall of the intestine and enters the blood stream. It eventually reaches the lungs and is expired together with carbon dioxide, water etc. Everyone, has a background level of hydrogen in their breath. If the bacteria in the gut get extra 'food' they will produce more hydrogen which will be detected in the breath. Bacteria get more 'food' whenever the body is unable to absorb nutrients. In this test patients are given a special test meal in the evening and instructed to fast for 12 hours. After this period, end-expiratory breath samples are collected every half-hour for 2 hours. The next day a control is carried out: the fasting patient is given 50 grams of glucose in 250 mL of water and the 2 hour excretion of breath hydrogen is measured. A high fasting breath hydrogen level and an increase of 12 parts per million = positive for bacterial overgrowth. [☛ Glycolic acid breath test].

Calcitonin

Reference range: < 150 ng/L {<150 pg/mL}

Calcitonin is produced in the parafollicular ('C-cells') of the thyroid gland [☞ Thyroxine, Triiodothyronine] in response to a high concentration of blood calcium ions (Ca^{2+}). The exact role and significance of calcitonin in human calcium homeostasis is not clear at present (patient who have had their thyroids removed do not lose control of calcium homeostasis). In general calcitonin lowers calcium concentration by (1) opposing bone resorption; (2) increasing renal excretion.

Calcium

Reference range (total, serum): 2.1-2.6 mmol/L {8.4-10.2 mg/dL}
Reference range (ionized): 1.12-1.23 mmol/L {4.48-4.92 mg/dL}

Total serum calcium is made up of: calcium ions which are: (1) bound to albumin and other proteins (40% of total); or (2) bound to smaller anions (15%) such as bicarbonate, citrate and phosphate; or (3) unbound, free ions (45%) a.k.a. 'ionized calcium'. It is *ionized* calcium (Ca^{2+}) that is the physiologically active entity. *Total* calcium concentrations do not necessarily always reflect ionized calcium levels: thus, measurement of total calcium can be misleading in certain circumstances. The concentration of <u>total</u> calcium in serum is affected by the concentration of albumin [☞ Albumin] in serum. Why? Ionized calcium (Ca^{2+}) binds (non-covalently) to albumin. The smaller the amount of albumin the greater the amount of unbound calcium and the greater the <u>measured</u> concentration of calcium appears to be. The opposite is true if the amount of albumin is increased. In addition as the concentration of hydrogen ions (H^+) increases (pH decreases [☞ pH]), i.e., acidosis [☞ acid-base], Ca^{2+} ions are displaced from their albumin binding sites and therefore the concentration of free Ca^{2+} increases. Remember that Ca^{2+} is the physiologically active entity. The opposite occurs if the H^+ concentration falls (increase in pH).

Calcium is essential for normal bone structure and growth and as a 'second messenger' in relaying surface receptor messages within the cytoplasm. Over 99% of total body calcium is in bone. Therefore, serum calcium measurements are NOT a good index of total body calcium. Uptake of calcium (intestine) is mediated by vitamin D Loss of calcium is 'fine tuned' by the distal tubules of the kidneys under the control of parathyroid hormone (PTH) [☛ Parathyroid Hormone (PTH)] PTH increases reabsorption of calcium. Calcitonin [☛ Calcitonin] decreases the concentration of calcium in serum. Note also that the concentration of calcium is also influenced by pH (see above) and the amount of bone resorption (iconoclast cells; stimulated by PTH {vitamin is also required}) versus the amount of bone building (iconoblast cells). This dynamic process is going on all the time. In health the net effect is calcium homeostasis.

Increased calcium can be caused by: cancer (common observation), primary hyperparathyroidism, drugs, multiple myeloma, hyperthyroidism and sarcoid.

Decreased calcium can be caused by: kidney failure, hypoparathyroidism, magnesium deficiency (magnesium is required for release of PTH).

Carcinoembryonic antigen (CEA)

Reference range: < 5 µg/L {<5 ng/mL}

CEA is most useful for monitoring cancer treatment. CEA is NOT diagnostic of cancer (not a screening test). CEA may be increased in smokers and ex-smokers, cancers of the colon, rectum, pancreas, lung etc. However, it may also be increased in non-malignant conditions such as emphysema, colitis and alcoholic cirrhosis.

CEA, as its name implies, is a normal constituent of embryonic blood. It is synthesized by the intestinal cells of the fetus. In normal healthy adults levels of CEA are very small (< 5 µg/L).

The 'carcino' part of CEA is derived from the Greek word, 'karkinos', meaning, 'a crab'. In ancient times, the term' 'karkinos', was used to describe any hard, non-healing ulcer. By the nineteenth century the term, 'carcinoma' was restricted to its current meaning, i.e., malignant neoplasms.

The 'embryonic' part of CEA is derived from the Greek word, 'embryon' meaning, 'the fruit of the womb'. 'Embron' is actually a compound of two words, 'en' meaning 'in' and, 'bryo' meaning, 'to cause to burst forth'.

'Antigen' is yet another example of invented 'Chemspeak' or 'Iatrish'. An antigen is anything that stimulates antibodies. It is derived from the Greek words, 'anti' meaning, 'against' and the word 'gennao' meaning, 'I produce'. Thus, antigens 'produce' (stimulate, generate) antibodies.

Casts

Not to be confused with (East) Indian class structures, iron objects or magical spells. Urinary casts are microscopic bodies found in some samples. They are often tube-shaped with a translucent sheath (made up mainly of Tamm-Horsfall glycoprotein [Tamm-Horsfall Protein)) encasing various elements such as cells, etc. The main types are:

Cast	Description	Interpretation
Hyaline	usually transparent few or no inclusions	1 or 2 per low powered field under microscope = normal
Granularx	transparent sheath containing granules	may be normal or pathologic
RBC (red blood cell; erythrocyte	transparent sheath containing rbcs	pathologic: severe injury to the glomerulus or (rare) transtubular bleeding
WBC (white blood cell)	transparent sheath containing wbcs	pathologic: infection especially pyelonephritis; glomerular disease.

[☛ Dipsticks, Urinalysis]

CCK-secretin Test

[☞ Cholecystokinin-secretin test]

CEA

[☞ Carcinoembryonic antigen]

Cerebrospinal Fluid (CSF)

Glucose: reference range: 2.8-4.2 mmol/L {51-76 mg/dL}
Protein: reference range: 250-550 mg/L {25-55 mg/dL}

CSF is a fluid produced in the ventricle of the brain. It fills the ventricles and bathes the brain and spinal chord; and is replaced 4-5 times per days. Total volume: 130-150 mL. CSF is a modified ultrafiltrate of plasma. Essential differences between CSF and plasma include: CSF protein concentration are $1/100^{th}$ those of plasma; CSF calcium levels are lower than serum (close to those of ionized calcium in plasma); CSF glucose levels are 10-20 % less than those in plasma. HCO_3 diffuses only slowly into the CSF from the plasma: thus, there is often a lag or delay between acid-base [☞ Acid-Base] changes in the plasma and changes in the CSF. Note that in contrast, CO_2 diffuses rapidly between CSF and plasma.

The main tests of CSF are: microbiological (infectious organisms, cell counts and types etc) and biochemical (protein and glucose). The color of the CSF is significant: clear/colorless = normal; bright red: blood present (poor spinal tap? subarachnoid hemorrhage?); yellow ('xanthochromic' {from the Greek word, 'xanthos' meaning, 'yellow'}) = a hemorrhage that has occurred in the last few days (color = bilirubin and other breakdown products of hemoglobin [☞ Bilirubin]).

(1) Protein: increased: infection, blood contamination, chronic CNS disorders (multiple sclerosis; TB, syphilis, Froin's Syndrome {blockage of spinal fluid flow through canal from, e.g., prolapsed vertebral disc, tumor, infection etc}).
(2) Glucose: decreased: infection (bacterial and leukocytes); hypoglycemia. Increased: hyperglycemia.

'Cerebrum' is the Latin word for, 'brain'.

Chemspeak

[☞ Iatrish]

Chiropractor

One who takes Clinical 'Biochemisery' courses inspite of him/ herself. From the Greek, 'cheir', meaning, 'a hand' and 'praktikos' meaning, 'fit for doing'; emphasizing their 'manipulative' nature (professionally)!

Chloride

Serum reference range: 95-106 mmol/L {95-106 mmol/L}
Urine reference range:110-250 mmol/day {110-250 mmol/day} (varies widely, depending upon chloride intake)

One of the electrolytes (sodium, potassium, chloride and bicarbonate). One of the major extracellular anions [☞ Anion Gap]. When bicarbonate (HCO_3) leaves red blood cells, chloride (Cl) enters in order to maintain an overall neutral charge. This is called the 'chloride shift'. There is a rough inverse relationship between bicarbonate and chloride. This is observed particularly in acid-base disturbances. [☞ Acid-Base].

Urinary chloride is not routinely available in most labs. However, it can be useful in the differential diagnosis of persistent metabolic alkalosis. The most common causes, of which, are: loss of gastric juice (HCl) and diuretic induced. Patients with these conditions will respond to chloride administration.

The word, 'chlorine' is derived from the Greek word, 'chloros', meaning 'green'. (Chlorine gas is green colored).

Cholecystokinin-Secretin Test

Reference range: total volume of secretions > 150mL
peak bicarbonate >60mL
total bicarbonate post secretin >5.7mmol
peak tryptic activity >30I.U. mL
total tryptic activity after CCK >2200 I.U.

A direct test of pancreatic *exo*crine function. The pancreas is both an *exo*crine and an *endo*crine gland [☛ Glucose, Glucose Tolerance Test;]. The pancreas secretes bicarbonate and other fluids in response to stimulation by the hormone secretin. Secretin is itself secreted by special cells located in the small intestine. Cholecystokinin (CCK) is stimulates the secretion of pancreatic enzymes such as lipase, amylase and chymotrypsin (these enzymes breakdown fats, carbohydrates and proteins respectively from large polymeric molecule into smaller units). CCK is itself secreted from special cells in the small intestine. The test is conducted in two main parts. Part 1 tests secretin function and part 2 tests CCK function. As in most of these types of gastrointestinal function tests, the patient must fast overnight before starting the test. A double-lumen tube is inserted into the patient's G.I. tract. One tube stays in the stomach and is used to remove gastric acid throughout the test. No gastric acid can be allowed to enter the duodenum because it would neutralize any bicarbonate that was present. Remember, bicarbonate is secreted by the pancreas and we are trying to measure how much bicarbonate the pancreas can secrete. If gastric acid is allowed to neutralize bicarbonate in the duodenum we will get a misleadingly low value for the amount of bicarbonate produced. The other tube is placed inside the duodenum close to the sphincter of Oddi (this is the inlet where the pancreas is connected to the small intestine). Duodenal juices are collected and stored on ice (many of the pancreatic enzymes are labile. Lowering the temperature helps them to 'live' longer in the collection vessel). Next a synthetic version of secretin is administered, i.v. and duodenal juices collected for about 30 minutes. There is then a 30 minute interval. Next a synthetic version of CCK is administered, i.v., and duodenal juice is collected as before.

The total volume of juice; the peak bicarbonate concentration; the total amount of bicarbonate secreted and the activity of the various enzymes secreted give information about general pancreatic sufficiency.

Cholesterol

Reference range: see below

The good, the bad and the confusing: 'Good' cholesterol is High Density Lipoprotein (HDL). Increased levels of HDL are believed to be beneficial. 'Bad' cholesterol is Low Density Lipoprotein (LDL) .

Despite the advent of low-cholesterol everything (except for low cholesterol income tax forms) the whole question of cholesterol and arteriosclerosis is still the subject of much debate. Each country has its own guidelines for testing—even within for example, Canada each province has different guidelines! One set of guidelines (Canadian Society of Clinical Chemists) is:

(1) Measure total cholesterol on any adult (20-65 years old) with at least 2 risk factors (risk factors = male; family history of premature coronary heart disease (CHD); hypertension; diabetes mellitus; cigarette smoker; obese; lack of exercise).

Desired total cholesterol: < 5.2 mmol/L {< 200 mg/dL}
Borderline total cholesterol: 5.2-6.2 mmol/L {200-240 mg/dL}
High total cholesterol: >6.2 {>240 mg/dL}

(2) For children > 5 years old or patients with a family history of CHD, measure: Total cholesterol, triglycerides [☛ Triglycerides], HDL and LDL.

Desired total cholesterol: < 4.6 mmol/L {<178 mg/dL}
Borderline total cholesterol: 4.6-5.7 mmol/L {178-220 mg/dL}
High total cholesterol: >5.7 mmol/L {> 220 mg/dL}

Total cholesterol may be increased in: lipoproteinemias type II, III and V; cholestasis; nephrotic syndrome; hypothyroidism; oral contraceptives; pregnancy etc.

Total cholesterol may be decreased in: liver disease; malabsorption; malnutrition; hyperthyroidism; abetalipoproteinemia etc.

'Cholesterol' was first discovered in gall stones and was thought to be bile that had solidified. Thus, it was named from the Greek words, 'chole' meaning 'bile' and 'steros' meaning 'solid'. Not bad for 'Chemspeak'!). This gave the original name, 'cholesterin' which was later changed to 'cholesterol' to emphasize that it was an alcohol.

CK

[☛ Creatine Phosphokinase]

Clinical Chemist

A strange species. Half-(wo)man, half-scientist, half-manager. Poor at adding up fractions. Generally, a person holding a doctorate or higher degree, trained and certified to practice Clinical Chemistry.

Clinical Chemistry

A strange and mysterious, mad pursuit. The analysis of body substances (blood, serum, plasma, urine, CSF, amniotic fluid, transudates, exudates, feces, intestinal liquids, breath gases etc). Interpretation of data obtained from these analysis as an aid to diagnosis, prognosis and treatment.

People are—medically speaking—black boxes. Much time, effort and money is expended on finding out what is going on inside the box. Clinical diagnosis is essentially the extraction of relevant information from the box. The information should be as timely, precise and accurate as is possible. Of course with the signal comes much noise. Years of training and experience are necessary in order to filter out the wanted signal from the background noise.

We carry a formidable array of tools around with us all the time. We have a *voice* which we can use to 'probe' patients—asking for a history. We have *ears* (twice as many as the number of mouths on average: is there a message here?!) to listen to his or her replies; and to listen to lungs, pulses and borborygmi etc. We have stereoscopic full-colour (most of us!) *eyes* to gather light waves reflected from the patient so that we can see rashes, spots, and all manner of signs. We have *hands* that we can palpate and percuss. We have *smell* that we might detect acetonic breath or halitosis. At one time we even used *taste*—ye olde urinalysis often included a tasting of patient's urine detecting such conditions as diabetes mellitus. Coordinating and interpreting all these sensors is the world's greatest analog computer; the human brain.

'Personkind' is a tool maker and user. Medicine has been quick to embrace tools—*extensions of our senses*— of many kinds to aid in clinical diagnosis. From simple, low-tech tools such as the reflex hammer; the stethoscope and mercury thermometers through 'medium-tech' tools such as the ophthalmoscope to modern, high-tech, tools. Once the electro-weak force was discovered [electricity, magnetism and radioactivity are all different manifestations of the same basic force—just as ice, steam and water are different manifestations of H_2O) things really started to take off:

the whole of the electromagnetic spectrum became available both for information gathering and for therapy. The discovery of X-rays in 1896 by Roentgen ushered in the new era of technological medicine. Since that time we have called into service ever wider expanses of the electromagnetic spectrum and the electro-weak force: from Magnetic Resonance Imaging (MRI) to Positron Emission Tomography, from infra-red sensing to ultra-violet therapy and from neutron bombardment to technetium ^{99} scanning.

Conn's Syndrome

Primary hyperaldosteronism, i.e., an increase in the amount of aldosterone [☛ Aldosterone] caused by a tumor in the cortex of the adrenal gland. Typical symptoms and signs: hypertension, hypokalemia, alkalosis and low plasma renin.

[☛ Addison's Disease, Cushing's Syndrome]

Cortisol

Reference range: 160-660 nmol/L {6-24 µg/dL} (08:00 hrs)

Cortisol is released by the adrenal cortex of the adrenal glands into the blood stream in response to ACTH (released from the anterior pituitary) [ACTH]. As the concentration of cortisol increases there is a negative feedback effect by cortisol on the release of ACTH. CRF, ACTH and cortisol all exhibit diurnal variation. Random samples must therefore be collected at a consistent time of day, preferably when levels are expected to be at or close to their peak (08:00 a.m.) By 24:00 hrs levels will be approximately one-half of those seen at 08:00 a.m. ACTH is useful in the differential diagnosis of Cushing's Syndrome [☛ Cushing's Syndrome, Addison's Disease, Conn's Syndrome].

'Cortex' is the Latin word for 'bark or shell'. This where we get the names adrenal cortex, cerebral cortex etc: all referring to the outer covering or part of the organ in question. Cortisol, corticosteroid, etc. all refer to substances associated with the cortex (adrenal).

CPK

[☞ Creatine Phosphokinase]

Creatine

Reference range: 13-53 μmol/L {0.17-0.70 mg/dL} (male), 27-71 μmol/L {0.35-0.93mg/dL} (female).

Creatine is synthesised mainly in the liver. It is transported to the muscles where it is phosphorylated:

$$\text{creatine} + \text{phosphate} \xrightarrow{\text{creatine phoshokinase}} \text{creatine phosphate}$$

Creatine phosphate is an energy store for muscle activity. The amount of creatine in the blood in healthy patients depends upon muscle mass and explains why the reference range is gender specific.

Creatine and <u>creatinine</u> are yet another example of 'chemspeak'. Paranoid students, might suspect that these two words were deliberately made to look and sound almost the same, in order to increase their confusion. Creatinine is the anhydride of creatine. Both names are derived from the Greek word, 'kreas' meaning, 'meat'.

Creatine Phosphokinase (CPK or CK)

Reference range: Total CK: 0-100 IU/L {0-100 IU/L}. CK-3 (94-100%), CK-2 (0-6%), CK-1 (0%).

CK (also called CPK) is found in high concentrations in: muscle, brain and heart. It exists as different isoenzymes [☞ Isoenzyme]. The 'old' nomenclature for CK isoenzymes should be replaced with the 'new':

Old	New	
CK-BB	= CK1	brain
CK-MB	= CK2	heart increases in acute myocardial infarction
CK-MM	= CK3	muscle

CK-MB increases within 2-12 hours, peaks at 12-40 hours then returns to reference range values within 24-72 hours. CK-MB increases in other conditions (e.g., Duchenne's muscular dystrophy, Reye's syndrome, strenuous exercise such as marathon runners etc.).

creatine phoshokinase
creatine + phosphate ➔ creatine phosphate

Using electrophoresis or specific antibodies the three isoenzymes can be measured and compared with the total CK levels.

CK-MB, etc. have no connection with FM radio stations of similar sounding names.

Creatinine

Reference range: 53-106 μmol/L {0.6-1.2} (male), 44-97 μmol/L {0.5-1.1 mg/dL} (female)

Creatinine (not to be confused with <u>creatine</u> [☛ Creatine) is useful in checking on kidney function [Creatinine Clearance, Urea]

Creat*inine* is formed from crea*tine*, in the muscles. The amount of creatinine in the blood in healthy patients depends upon muscle mass and explains why the reference range is gender specific.

CSF

[☛ Cerebrospinal Fluid]

Cushing's Disease

[☛ Cushing's Syndrome]

Cushing's Syndrome

Not to be confused with Cushing's <u>disease</u> [☞ Syndrome]. Cushing's <u>syndrome</u> is caused by excess cortisol (hypercortisolism). This can be due to several different causes: (1) cancer of the adrenal cortex → excess cortisol secretion = 'primary' . (2) excess production of ACTH [☞ Adreno-corticotropic Hormone (ACTH)], by the pituitary = 'secondary' = Cushing's <u>Disease</u>. (3) ectopic secretion of ACTH by tumors (e.g., carcinoma of the bronchus). (4) Iatrogenic, i.e., corticosteroid treatment.

Symptoms and signs may include: truncal obesity; 'moon' face; buffalo hump (altered fat distribution); thin skin; easily bruised; hirsutism, skin pigmentation (only if ACTH is elevated {not seen in primary Cushing's}; hypertension; glucose intolerance; menstrual irregularities; psychiatric problems {depression; mania; euphoria}).

Cynical Chemist

A jaded Clinical Chemist. [☞ Clinical Chemist]

Darlington

A town in the North-East of England of 90,000 people. The first passenger steam railway ran from Darlington to Stockton in 1825. The Darlingtonocentric Theory of Creation, postulates that the universe as we now know it was accidentally created during a random quantum fluctuation in a pint of 'Old Peculiar' at the 'Chequers Inn' in Darlington. This theory is not widely accepted outside of Darlington.

Diabetes Insipidus

Not to be confused with diabetes mellitus [☛ Diabetes Mellitus]

Due either to a deficiency of anti-diuretic hormone (ADH) (also known as 'vasopressin') [☛ Anti-diuretic Hormone]. This is known as neurogenic diabetes insipidus. Or to dysfunctional receptors for ADH in the collecting ducts of the kidney (nephrogenic diabetes insipidus). The two types can be differentiated by giving the patient ADH.

Patients with diabetes insipidus are unable to retain water and therefore have polyuria (assuming sufficient intake of water), very dilute urine and excessive thirst.

'Diabetes' is from the Greek, 'dia' and 'baino', which combined mean, 'a syphon'. Patients with this disease are indeed 'fountain-like' with respect to their urine output. The 'insipidus' refers to the fact that the dilute, pale urine is tasteless (compared with the sweet taste of urine from patients with diabetes mellitus). [☛ Diabetes Mellitus, Urinalysis].

Diabetes Mellitus

Not to be confused with diabetes insipidus [☛ Diabetes Insipidus]

Diagnosis of diabetes mellitus is dependent on clinical evaluation in addition to the Glucose Tolerance Test [☛ Glucose Tolerance Test]. There are two types of this 'disease': Type 1 = Insulin Dependent Diabetes Mellitus (IDDM) and type 2 = Non-insulin Dependent Diabetes Mellitus (NIDDM). IDDM generally affects people under 30 years who become dependent on insulin injection for survival. NIDDM affects older people and may be associated with obesity and aging. As the name implies patients are not dependent on insulin injections for survival.

Symptoms and signs of diabetes mellitus may include: hyperglycemia, glucosuria, thirst, polyuria. Long-term complications: damage to kidney, nervous system, retinae and blood vessels.

'Diabetes' is from the Greek, 'dia' and 'baino', which combined mean, 'a syphon'. Patients with this disease are indeed 'fountain-like' with respect to their urine output.' 'Mellitus' is from the Latin meaning 'sweetened with honey' (a reference to the sweet tasting of urine of diabetes mellitus patients. A dangerous if somewhat low-tech test!).

Dipstick

Urine is usually, more or less, free. Unlike serum which we usually have to go in and get via venepuncture, urine is a freebie. It carries within it information about what is going on inside the body that just excreted it. Dipsticks are sticks with several small pads attached. Each pad contains specific chemicals that react with certain substances present in the urine. The pads change color when this happens. The degree of color change is related to the amount of substance present in the urine allowing semi-quantitative estimates of concentration or number. Dipsticks allow as many as 9 different tests to be performed all at once. Color change is usually read within 1 to two minutes. This can be done by eye (the color of the pad is compared with a color reference chart) or by an automatic dipstick reader. Great care must be taken with the interpretation of these tests. Urine samples must be fresh (< 2 hours old). Dipsticks may contain some or all of the following test pads:

Test	Associated disease (examples)
pH	acidosis; alkalosis;
protein	nephrotic syndrome
glucose	diabetes mellitus
pecific gravity	diabetes insipidus; renal failure
ketones	diabetic ketoacidosis; toxins
bilirubin	liver disease:
leukocytes (wbcs)	blood ection of tract

[☛ Urine, Casts]

Disappointment

[☛ Hormone]

Drugs

A drug is any substance that when injected into a rat will produce a scientific paper.

Name	Synonym 1	Synonym 2	Synonym 3
Acetaminophen	Distalgesic	Paracetamol	Tylenol
Acetylsalicylic acid	Aspirin		
Amikacin	Amikin		
Amikin	Amikacin		
Amiodarone	Cordarone		
Amitriptyline	Tryptizol		
Aspirin	Acetylsalicylic Acid		
Bruffen	Ibuprofen		
Carbamazepine	Tegretol		
Chloramphenicol	Chlormycetin		
Chlormycetin	Chloramphenicol		
Cordarone	Amiodarone		
Cremostrep	Streptotriad	Streptomycin	Sulphamagna
Depakene	Epilim	Valproic Acid	
Digitaline	Digitoxin	Nativel	
Digitoxin	Digitaline	Nativel	
Disopyramide	Norpace	Rhythmodan	
Distalgesic	Acetaminophen	Paracetamol	Tylenol
Epilim	Valproic Acid	Depakene	
Ethosuximide	Zarontin		
Flecainide	Tambocor		
Gentamicin	Genticin	Genitsone	
Genticin	Gentamicin	Gentisone	
Gentisone	Genticin	Gentamicin	
Ibuprofen	Bruffen		

Imipramine	Tofranil		
Inderal	Propranolol		
Kanamycin	Kantrex		
Kantrex	Kanamycin		
Kindin	Quinidine		
Lidocaine	Xylocaine		
Methotrexate	Methotrexate LPF	Mexate	
Methotrexate LPF	Methotrexate	Mexate	
Mexate	Methotrexate LPF	Methotrexate	
Nativel	Digitaline	Digitoxin	
Nebcin	Tobramycin		
Netilmicin	Netromycin	Netilin	
Norpace	Disopyramide	Rhythmodan	
Paracetamol	Distalgesic	Acetaminophen	Tylenol
Phenytoin	Dilantin	Epanutin	
Primidone	Mysoline		
Procainamide	Pronestyl		
Pronestyl	Procainamide		
Propranolol	Inderal		
Quinidine	Kindin		
Rhythmodan	Norpace	Disopyramide	
Streptomycin	Streptotriad	Cremostrep	Sulphamagna
Streptotriad	Streptomycin	Cremostrep	Sulphamagna
Sulphamagna	Cremostrep	Streptotriad	Streptomycin
Tambocor	Flecainide		
Tegretol	Carbamazepine		
Tobramycin	Nebcin		
Tocainide	Tonocard		
Tofranil	Imipramine		
Tonocard	Tocainide		
Tryptizol	Amitriptyline		
Tylenol	Paracetamol	Distalgesic	Acetaminophen
Valproic Acid0	Empilim	Depakene	
Vancocin	Vancomycin		
Vancomycin	Vancocin		
Xylocaine	Lidocaine		
Zarontin	Ethosuximide		

Education

Reference range: incredibly dull to brilliantly inspiring with a standard deviation of so-so.

You've heard of a triple-E senate: *e*xpensive, *e*xtinct and *E*-relevant, well, I think that there should also be triple-E education. When the BBC (British Broadcasting Corporation) was founded, its mandate (now changed to a personate) was to Entertain, Educate and Enlighten. This makes good sense. The idea is to entertain and thereby gain the student's/audience's attention and then to insidiously slip in some education. On a good day, with a clear sky and the wind at your back it is even possible to enlighten. Switching the lights back on in the lecture theatre, after the last slide, doesn't count.

Efficiency

A measure of how many tests overall were correct: the sum of all true positives (TP) and all true negatives (TN) divided by the total number of test performed = TP + TN + FP + FN (where FP = false positives and FN = false negatives).

$$\text{Efficiency} = \frac{TP + TN}{TP+TN+FP+FN}$$

Electrolytes

[☞ **Sodium, Potassium, Chloride and Bicarbonate**].

One of the most frequently ordered set of tests. Essential indicators of general health status of patient. [☞ Anion Gap]

Electrophoresis

Separation of substances in solution by passing an electrical field through them. Modern electrophoretic systems use acetate or agar gel as a matrix on which to apply the sample. The matrix is placed in a buffer and subjected to an electrical field for a time. Molecules are separated on the basis of charge and size. The method is used for example to separate, isoenzymes of alkaline phosphatase (ALP) and of Creatine Kinase (CK) [☛ Alkaline Phosphatase, Creatine Phosphokinase].

Ethanol (alcohol)

Reference range: 0

The word 'alcohol' is derived from the Arabic words, 'al' meaning, 'the' and 'koh l meaning, 'fine powder'. The first 'al koh l' was a kind of Arabic mascara: made of finely ground antimony it was applied to the eyes as a kind of make-up. The term took on the general meaning of anything that could ground up to make a very fine powder such that no residual grains could be seen. Oddly enough, 'no residue' came to be associated with any volatile substance that evaporated and left no residue.

Examinations

Reference range: varies widely from 'failed miserably' {brain stem dead} to '110 out of 100' {outstanding}

A way of finding out about things or people or understanding. Examining a patient both orally and physically can yield much useful clinical information. Examining a student's understanding of material covered in lectures or whatever is very difficult to do well. Essays and short answers tend to invite massive 'brain-dumps' of anything remembered about the last 10 years. They also favour people who can write quickly, clearly and legibly, none of which may be under assessment at the time. Multiple choice questions can avoid some of these problems but are difficult to set. The best answer to this dilemma is to use the Vulcan mind probe. [☛ Education]

Fecal Fat

Reference range: < 7 g/day {< 7 g/day}

Patients on a balanced diet should excrete less than 7 g/day of fat in their feces. Collection of stools for a 72 hour period allows the measurement of fecal fat excretion. Increased excretion (steatorrhea) may be seen in: pancreatic disease/blockage; bile salt deficiency/blockage and impaired intestinal absorption.

Getting patients to actually play the game according to your specified instructions is often difficult. Patients have been known to pass their stools into the toilet bowl and then fish them out (unfortunately also bringing with them bleach, urine, blue, pink or whatever other colored 'sanitizer' dyes etc) and put them into the collection vessel.

The word, 'feces' is Latin for 'dregs or sediment'. Until the seventeenth century 'feces' had the same general meaning in English until it took on its modern, less fluid, meaning.

Fecal Occult Blood

Reference range: none detectable

Not a bizarre secret ritual performed by Clinical Chemists. The term 'occult' is used in the sense of being 'hidden'. Colorectal cancer commonly leads to intestinal bleeding resulting in a positive occult blood test. False positives can be caused by, e.g., bleeding gums, meat in diet, etc.

Ferritin

Reference range: 20-250 µg/L {20-250 ng/mL} (male)
Reference range: 10-100 µg/L {10-100 ng/mL} (female)

Ferritin is the major iron storage protein [☞ Total Iron Binding Capacity] It is found in liver, spleen and bone marrow reticuloendothelial cells. A small amount also occurs in serum. Decreased levels may be seen in: iron deficiency. Reduced iron stores are the only cause of reduced ferritin levels; and is the best single biochemical test of body iron stores [☞ Iron]. Increased amounts may be seen in iron overload diseases (e.g., hemochromatosis). There are many reasons for increased ferritin levels, other than due increased iron: acute liver diseases, leukemias, lymphomas, carcinomas, acute and chronic infections (ferritin is an acute phase protein).

Iron containing compounds, e.g., hemoglobin, are named from the Latin word, 'ferrum' meaning 'iron'. Disease states arising from imbalance of iron homeostasis, e.g., hemosiderosis, are named from the Greek word for iron: 'sideros'. This inconsistency in 'Iatrish' is not a conspiracy against students. Cynical students may suspect the sinister hand of the *International Committee for Giving Things Totally Confusing and/or Unpronouncable Names That Have Little or Nothing to Do With Anything'.*

Fluorescein-Dilaurate Test

Reference range: <u>*fluorescein excreted on day 1*</u> x 100
$\qquad\qquad$ *fluorescein excreted on day 2*
$\qquad\qquad$ *Normal pancreatic function > 30%*
$\qquad\qquad$ *Pancreatic insufficiency < 20%*

An *in*direct test of pancreatic function. Fluorescein fluoresces and can be detected in urine. In this test a fasting patient is given fluorescein dilaurate. This is man-made: fluorescein attached to a fat (dilaurate). The patient drinks a solution containing the fluorescein dilaurate which eventually reaches the duodenum. If the patient's pancreas is functioning normally then it will secrete the enzyme lipase. Lipase breaks down fats into smaller units. It cleaves the fluorescein from the dilaurate. The fluorescein is absorbed by the cells of the small intestine and it enters the hepatic portal vein and arrives in the liver. It then enters the general circulation and is eventually excreted via the kidneys and appears in the urine. The fluorescein present in the urine is detected by illuminating the urine and measuring the amount of fluorescent light emitted. If none, or very small amounts of fluorescein are detected in the urine there are three possible reasons why:
$\qquad$ (1) the patient lacks lipase (or the lipase is not functioning);
$\qquad$ (2) the patient is unable to absorb fluorescein and transport it to the liver; or
$\qquad$ (3) the patient is unable to excrete fluorescein due to renal disease or urinary tract obstruction.

Causes (2) and (3) can be eliminated by performing a control test: the day after the initial test (fluorescein-dilaurate described above) the patient is given fluorescein in solution. It is not attached to anything. If fluorescein appears in the urine, it means that the patient *can* absorb fluorescein and transport it to the liver *and* excrete it via the kidneys. Thus, you can now conclude that explanation (1) is the correct one , i.e., the patient lacks lipase (or the lipase is not functioning). This is usually due to pancreatic insufficiency but it may also be due to excess acid entering the duodenum (lipase works best in an alkali environment) from the stomach as can occur in hypergastrinemic conditions such as the Zollinger-Ellison syndrome. [☛ Pentagastrin test; Zollinger-Ellison syndrome].

Follicle Stimulating Hormone (FSH)

Reference range (male): 4-25 IU/L {4-25 mIU/mL}
Reference range (female, mid-cycle peak): 10-90 IU/L {10-90 mIU/L}
Reference range (female, post-menopausal):40-250 IU/L {40-250 mIU/mL}
Reference range (female, pregnant): undetectable or very low

In the female, FSH governs development of the ovarian follicle (At last! A name that describes a function!) and of the ova. In the male, FSH governs testosterone [☛ Testosterone] and sperm cell development (in conjunction with luteinizing hormone (LH) [Luteinizing Hormone (LH)].

In females, FSH levels increase at age 10-11 years (About 2 years later for males). During the menstrual cycle FSH levels increase abruptly (2 x base levels) at about day 15. Ovulation occurs 1-2 days later and FSH levels fall to base levels by day 20. Post-menopausal women have relatively high levels of FSH (and LH) because these women are no longer ovulating, i.e., there is no longer any negative feedback suppression of FSH and LH.

Free T₃
[☛ Triiodothyronine]

Free T₄
[☛ Thyroxine]

FSH
[☛ Follicle Stimulating Hormone]

Gamma glutamyltransferase

Reference range: 0-30 IU/L {0-30 IU/L}

Also known as gamma-glutamyltranspeptidase. Increases following almost any insult to the liver (except for sarcasm and hyperbole). Is very sensitive but not very specific. Is useful in monitoring abstinence in chronic alcoholics.

GH

[☛ Growth Hormone]

GGT

[see Gammaglutamyl Transferase]

Glomerular Filtration Rate (GFR)

1-2 mL/s

GFR is an index of the kidney's ability to filter blood. The best test for GRF is to administer inulin (a plant carbohydrate, not to be confused with insulin), intravenously. Inulin is not metabolized, reabsorbed or secreted. It is only filtered. By measuring the amount of inulin 'cleared' from the serum in a given amount of time, it is possible to calculate the GFR:

$$\frac{[\text{Inulin}]\ \text{urine}}{[\text{Inulin}]\ \text{serum}} \times \text{volume of urine/second}$$

Inulin must be administered i.v., and requires a patient visit to a physician. Instead of using an exogenous marker such as inulin we can use a naturally occurring endogenous marker such as creatinine [☛ Creatinine].

Creatinine levels are more or less constant (<u>but</u> vary with: muscle mass, diet {meat contains creatinine}, whether pregnant, severe exercise etc). Unfortunately creatinine is secreted in addition to being filtered. Therefore, creatinine that appears in the urine is not entirely due to filtration and creatinine clearance will be higher than the true GFR. To calculate creatinine clearance, substitute creatinine in the above equation. Other difficulties include those inherent in collection of timed samples
[☛ Urine, 24 hours, Fecal Fat]

Glucose

Reference range: (fasting) 4-6 mmol/L {70-105 mg/dL}

One of the major energy sources for the body. Glucose is stored as glycogen. (A note on 'Chemspeak': Breakdown of glycogen to glucose is called 'glycogenolysis'. Breakdown of glucose, itself, is called 'glycolysis'. Synthesis of glucose from other substances is called 'gluconeogenesis').

A regular blood sample must be separated from cells quickly because cells 'eat' glucose and obviously glucose levels will fall giving a false result. This effect can be greatly reduced by collecting the blood into tubes that contain fluoride (an inhibitor of glycolysis). Blood glucose is often elevated in diabetes mellitus. This is NOT diagnostic of diabetes mellitus
[☛ Glucose Tolerance Test, Diabetes Mellitus].

Glucose detected in urine is strongly suggestive of pathological disease, e.g. diabetes mellitus (not to be confused with diabetes insipidus).
[☛ Diabetes Insipidus].

Glucose is yet another invented piece of 'Chemspeak'. This time it took an entire committee of the French Academy of Sciences to come up with it (July 16th, 1838: a date that shall live in infinity). Glucose is a very 'classy' word derived from not one but two classical languages: (1) from the Greek word, 'glukus', meaning 'sweet tasting' and (2) from the Latin 'osus'. The ending '-ose' became the standard 'Chemspeak' way of letting you know that a substance was a carbohydrate (glucose, fructose, lactose, pantyhose etc.).

Glucose Tolerance Test

Recommended test for diagnosis of diabetes mellitus (DM) in non-pregnant adults [☛ Diabetes Mellitus]. The psychological and socioeconomic consequences of being diagnosed as having DM are profound. This test must be prove positive on more than one occasion before a definitive diagnosis can be given. Decision criteria and target values may vary.

(1) If a <u>random</u> serum glucose is > 7.8 mmol/L {>141 mg/dL} then the tolerance test is warranted.
(2) Fast overnight (> 10 hours)
(3) Take a.m. blood sample to obtain a 'baseline' glucose concentration.
(4) Give oral glucose (75 g).
(5) Measure blood glucose at 1 hour
(6) Measure blood glucose at 2 hours

If the 2 hour sample and at least one other sample have glucose concentrations > 11.1 mmol/L {200 mg/dL} on more then one occasion; and the fasting glucose level was > 7.8 mmol/L {141 mg/dL}, then the patient has DM.

Note that it is not necessary to perform this test at all if: (a) at least two fasting glucose levels of > 7.8 mmol/L are obtained on separate occasions OR if a single <u>random</u> glucose of > 11.1 mmol/L {200 mg/dL} is observed together with classic symptoms [☛ Diabetes Mellitus].

Glycolic Acid Breath Test

Reference range: < 4 x 10^{-3}% total dose per mmol of $^{14}CO_2$ patients with active deconjugation >10 x 10^{-3}%

Test for bacterial deconjugation of bile salts. [☛ Breath Hydrogen Test for Bacterial Overgrowth, Fluorescein-dilaurate test, Cholecystokinin-secretin test]. In a normal healthy person bile salts are secreted from the liver and gall bladder into the small intestine. Bile salts are essential for micelle formation. Micelles are spherical groupings of ingested fats, bile salts and enzymes such as lipase (secreted by the pancreas). Most of the bile salts are reabsorbed in the ileum and recycled. This 'environmentally conscious' system is called the enterohepatic circulation.

In patients suffering from:
(1) disease of the ileum or
(2) surgical resection of the ileum or
(3) bacterial overgrowth of the *small* intestine
bile salt reabsorption is reduced or blocked completely. In conditions (1) and (2) the bile salts remain in the lumen of gut and travel down to the colon where resident bacteria deconjugate the bile salts. In condition (3)

break down occurs in the small intestine, i.e., the bacteria have 'travelled up to meet' the bile salts! In all three conditions the bacteria eventually break down the glycine part into CO_2. The CO_2 diffuses across the lining of the gut and enters the blood stream, eventually reaching the lungs where it is excreted in expired breath.

Following an overnight fast, patients are given an oral dose of a radioactively labelled synthetic bile salt: ^{14}C-glycolic acid which passes into the gut. The patient must fast for 2 more hours and then eat a meal. A baseline (pre-ingestion of ^{14}C-glycolic acid) breath sample is collected plus one 30 minutes post-ingestion of ^{14}C-glycolic acid and then hourly thereafter for 6 hours. If bacterial overgrowth of the small intestine has occurred $^{14}CO_2$ will appear in the breath early in the test. However, if the patient is suffering from either (1) disease of the ileum or (2) surgical resection of the ileum then $^{14}CO_2$ will appear in the breath late the test. This is because in conditions (1) and (2) the ^{14}C-glycolic acid must travel to the bacteria which are resident in colon whereas in condition (3) the bacteria have already 'travelled' to the small intestine and the ^{14}C-glycolic acid is available for deconjugation much sooner.

Growth Hormone

Reference range: males: 0-5 µg/L {0-5 ng/mL}
Reference range: females: 0-10 µg/L {0-10 ng/mL}

Also known as somatotropin. GH is secreted by the anterior pituitary in response to stimulation by the very sensibly named Growth Hormone Releasing Hormone. GH released is suppressed by somatostatin. GH acts upon the liver and certain other target organs facilitating somatomedin production. Somatomedins are responsible for growth stimulation and metabolic regulation. Excess GH can cause gigantism in children and acromegaly in adults. Deficiency of GH can cause dwarfism in children but appears to have no clinically discernible effect on adults.

GH secretion is episodic and pulsatile, peaking whilst asleep. As with many other hormones, GH function is best assessed by dynamic stress inducing tests, e.g., arginine infusion test; exercise test; insulin test).

GH is also known as somatotropin. 'Somato' is derived from the Greek word, 'soma' meaning 'a corpse or a body'. 'Tropin' is derived from the Greek word, 'tropos' meaning 'a turning'. In this sense GH is a hormone 'turned toward the body'.

HCO$_3$

[☞ Bicarbonate]

HDL Cholesterol

[☞Cholesterol]

Hormone

Substances secreted by endocrine glands that usually act somewhere else in the body: e.g.

> hypothalamus
>
> ⇓ *corticotrophin releasing hormone (CRH)*
>
> anterior
>
> pituitary
>
> ⇓ *adrenocorticotrophic hormone*
>
> adrenal glands
>
> ⇓ *cortisol*
>
> various cells

Hormone levels are usually controlled via feedback inhibition and/or direct inhibition of release.

Consider also...

Upon Not Hearing Gallo and Montagnier at the Gairdner Awards

The Disappointed Man slopes away from yet another un-event
And, after mentally noting those who shall have their Oxygen Licenses
  revoked
For failing to put Gallo-and-Montagnier in a place with seating for n + 1
+ me...
Home again, pondering the Biochemistry of Disappointment:

What phosphorylation or molecular switch
gives form to thought that "Life's a bitch?"
"Thoughts begat feelings." the psychiatrists say
But what is disappointment? What molecule per se?
Is it all just enzymes, substrates and NAD^+?
in my limbic system that make me feel bad;
some undiscovered organ, tissue or cell
that evolved to disappoint us, lest our lives go too well?

So with X-ray and CAT scan and MRI
And surreptitious soupsçi of sodium serendipiti
And years of accreting psychorrheic secreta
And secretly sifting excreta etcetera
I searched for the site of the Gland of Glum
From lumen to limbus, from lyra to lung.

Now, 20 years later—and nothing found.
The grants are no more and the bench space impound-
ed (by genetic engineers who only speak A.C.R.O.N.Y.M.).
My once-bright dream grown tarnished and dim
Of being a latter-day Banting and saving mankind
From much disappointment with an 'insulin for the mind'.

And so...

Disappointed I sit, aware deep within
Of the Zona Disappointa of that lost endocrine
Insidiously secreting pre-pro-disappointin.

I.Wilkinson ©1991, *The Medical Post,* February 26, 1991.

[☞ Addison's disease, Adrenocorticotropic Hormone, Aldosterone, Anti-
diuretic hormone, Calcitonin, Conn's syndrome, Cortisol, Cushing's
Disease, Diabetes insipidus, Diabetes mellitus, Glucose tolerance test,
Growth Hormone, Luteinizing Hormone, Parathyroid Hormone, Progesterone, Prolactin, Reverse triiodothyronine, Testosterone, Thyroid stimulating
hormone, Thyroxine, Triiodothyronine].

Humor

Reference range: very, very, very wide.

Unclear exactly what this is. Commander Data, an android on 'Star Trek The Next Generation' is extremely intelligent and has vast data banks on the endocrinic properties of resublimated thiotolamine, but cannot understand humor. Humor *may* be the unexpected juxtaposition of ideas such that a feeling of pleasure is evoked? There is a wide range of what most people consider funny: most humor involves in some way making fun of someone, some place or something. In Canada we are thankfully protected from humor by our constitution which is intermittently applied by whoever can afford to talk to the Supreme Court. Pressure groups such as *'Concerned Post-Pubescent But Pre-Menopausal Duck Billed Platypus Female Impersonators* (Marxist-Spencers Branch)' have managed to stamp out any kind of humor involving them and have government grants to prove it. We should appreciate the efforts of all such groups.

Iatrish

Doctorspeak. Closely related linguistically to 'Chemspeak'. A confused and confusing 'novoglot' of classical Greek and Latin, intermingled with English, French, German and Idiot. Whilst the disease itself is not yet fully understood, it has been described and tentatively named:

Primary Iatrogenic
Portmantologic Dysnomenclasia

No Roman or Greek
Was ever heard to speak
Of adiadochokinesia and so
Our medical societies
Spent hours out of piety
Inventing neo-classic lingo
Now there's no need to hammer

Every fine point of grammar
Nor to quote verbatim from Fowler
Or debate like cognoscenti
Ad infinitum (or infiniti?)
Over 'datas' versus 'datum' or 'data'
With congeni(men)tal block
Whether ergo propter hoc
Follows a posteriori—or not?

Thus, de facto, usage
Is replete with abusage
A more Iatrish than English novoglot
So when your diagnosis
Is "metabolic neutrosis"
(With no mention of a translator fee)
or when a phrase like "nephrotic nephrosis"
leads to aural stenosis
Then consider the poor patient's plea:

"Please say what you mean"
(Like in Alice's dream)
Stay away from the over complex
Please spare the confusion
Of these classical illusions
"Veritas absolutus sermo
Ac semper et simplex"

Ian Wilkinson

1990, First published in *The Medical Post,* reprinted in *The Naked Physician* Ed. Ron Charach, Quarry Press, 1990.

Iron

[☞ **Total Iron Binding Capacity**

Reference range: 10-30 μmol/L {56-168 μg/dL}

Isoenzyme or isozyme

One of a group of enzymes that catalyzes the same chemical reaction, but that differ in physical properties. Alkaline phosphatase (ALP) is an example: common isoenzymes of ALP include liver, intestine, bone and placental. Isoenzymes may be detected and measured by exploiting their physical and chemical differences using, e.g., electrophoresis [☞ Electrophoresis] or heat or acid inactivation techniques or isoenzyme specific antibodies [☞ Creatine Kinase].

Jargon

Reference range: From 'chronic metabolic neutrosis' (health, in English) to 'adopting an ongoing circumlocutious approach in order to maximize terminological inexactitude (lying, in English).

The fundamental unit of BS. Darlingtonocentrics believe that the entire cosmos (except of course for Betty Page) is composed of a number of fundamental particles such as the 'Boron' the fundamental unit of being bored out of your head in a lecture, or whatever; the 'Moron' the fundamental unit of government; the 'Luston': the fundamental initial attractive force between humanoids; the 'Klingon': the basic unit of someone who sticks to you and will not go away despite attempted pesticide; and the Wilkinson, which is defined elsewhere [☞ Wilkinson]. In science and medicine it is important to understand the meaning and correct use of jargon. Highly specific terms are precise tools that serve to avoid confusion or vagueness in describing a patient's condition. [☞ Sign, symptom, maldigestion, malabsorption and most of the rest of this 'Guide'].

It is vital that communication between health professionals be clear and precise. Despite the vast vocabulary of medical terms available to us, I often see in-patients, in our hospital described as, 'poorly', in the section of the chart that asks for clinical diagnosis. This is amazing; poorly people in a hospital. Whatever next?

Lactase

[☛ Lactose Tolerance Test]

Lactose Tolerance Test

Reference range: > 1.7 mmol/L (>31 mg/dL) over fasting glucose levels

In lactase deficiency the enzyme lactase, normally present in the lining of the small intestine, is absent or non-functional and the body is unable to convert the disaccharide (sugar) lactose to the monosaccharides galactose and glucose which would normally be absorbed by the body.

i.e., lactose ➔ galactose + glucose
         lactase

Thus, lactose passes into the large intestine where it is 'eaten' by the bacteria (resulting in increased hydrogen [☛ Breath Hydrogen Test for Bacterial Overgrowth]) that live there, rather than being absorbed into the blood stream. The test consists of drinking 50 grams of lactose in water and then measuring blood glucose 5, 10, 30, 60, 90 and 120 minutes later. In normal healthy patients the blood glucose should increase by > 1.7 mmol/L over fasting glucose levels (☛ Glucose). In lactase deficient patients the change is < 1.1 mmol/L. Note that similar tests can be performed for other disaccharidase deficiencies.

LDL Cholesterol

[☛ Cholesterol]

LH

[☞ Luteinizing Hormone]

Limerick

Reference range: from poetry to logorrhea

> There once was a Chiropractor
> Who swallowed a plastic protractor
> Though taking her D.C.
> She passed many degrees
> 180° to be more exact-er.

I. Wilkinson © 1991

> There once was a Doctor of Physic
> Who entered in contest, a limerick
> His self-referential style
> Made many a judge smile
> Though he lost his poetic license for this gimmick.

I. Wilkinson © 1987

Lipids

[☞ Cholesterol, Triglycerides]

Luteinizing Hormone (LH)

Reference range (male): 6-23 IU/L {6-23 mIU/mL}
Reference range (female, follicular phase): 5-30 IU/L {5-30 mIU/mL}
Reference range (female, post-menopausal): 30-200 IU/L {30-200 mIU/mL}

Also known as lutropin. LH is released (episodic and pulsatile) by the anterior pituitary. In males LH stimulates testosterone secretion by Leydig cells and spermatogenesis. In females LH stimulates ovulation and development of the corpus luteum [☞ Follicle Stimulating Hormone]. Levels of LH surge at about day 15 of the menstrual cycle rising to levels 6 x basal. LH is also increased in post-menopausal women due to lack of negative feedback by ovaries on anterior pituitary.

'Lutein' is derived from the Latin word, 'luteus' meaning 'mud-colored' (but taken to mean anything that is yellow in color. In yet another example of inconsistent Iatrish, the term 'xantho' (Greek origin) is also used to denote things that are yellow, e.g., 'xanthochromic' [Cerebrospinal Fluid]. The corpus luteum is the 'yellow body' that remains once the ovulation has taken place.

Lutropin

[☞ Luteinizing Hormone (LH)

Magnesium

Reference range: 0.7-1.1 mmol/L {1.3-2.1 mEq/L}

Magnesium is essential as a cofactor in many enzyme reactions and muscle contraction. Excess magnesium (e.g., renal failure) may induce cardiac arrest and/or respiratory paralysis. Hypomagnesia may be caused by: malabsorption, alcoholism, cirrhosis, diuretics, renal disease and chronic mineralocorticoid excess [☞ Conn's Syndrome].

'Magnesium' is named after the town of 'Magnesia', now in modern day Turkey, where the an ore containing magnesium was found in ancient times. This town also gave us the word, 'magnet' because another ore, found locally, had unusual 'navigational' and attractive/repulsive properties and became known as Magnesian stone. Not bad for one small town in Turkey (in the kingdom of Lydia in those days). Too bad Darlington did not have such ore (or awe)?—we could have navigated using a Darlingtonetic compass and Darlingtonium could have been element number 12 in the periodic table instead of boring old magnesium [☞ Darlington].

Malabsorption

Not to be confused with mal*digestion* [☞ Maldigestion]. The inability to take up nutrients from the lumen of the gastrointestinal tract into the body. Malabsorption may be a consequence of maldigestion or may be due to other causes for example: bacterial overgrowth [☞ Breath Test for Bacterial Overgrowth, Glycolic acid test].

The clinical features of malabsorption are due to two causes:

(1) Symptoms and signs (diarrhea, steatorrhea, abdominal discomfort, distension and flatulence) due to the retention of non-absorbed nutrients within the lumen of the gut. i.e., there is stuff in there that should not be there.

(2) Symptoms and signs (anemia, osteomalacia, edema, tendency to bleed, weight loss, failure to thrive in children and infants) due to decreased absorption of nutrients (iron, folate, vitamins B_{12}, A, D, E and K, proteins, fats, carbohydrates etc).

[☞ Cholecystokinin-secretin test; Pentagastrin test; Fluorescein-dilaurate test; Glycolic acid test; PABA test; Triolein breath test].

Maldigestion

Not to be confused with mal*absorption* [☞ Malabsorption] Maldigestion often causes malabsorption.

The inability to break down large molecules into smaller units and / or transport of these smaller units across the gut wall. Breakdown takes place in the lumen under the influence of pancreatic enzymes, bile acids and at the lining of the gastrointestinal tract under the influence of various enzymes.

For example:
    polysaccharides → disaccharides → monosaccharides, which can be absorbed
    triglycerides → fatty acids + glycerol, which can be absorbed
    polypeptides → peptides + amino acids, which can be absorbed

[☞ Cholecystokinin-secretin test, Pentagastrin test, Fluorescein-dilaurate test, Glycolic acid test, PABA test, Triolein breath test.]

Normal

[☞ Reference Range]

Osmolality

Reference range (serum): 275-295 mOsmol/kg {275-295 mOsmol/kg}
Reference range (urine): 50-1,400 mOsmol/kg {50-1,400 mOsmol/kg}

Not to be confused with osmolarity. Osmolality is the number of osmols per <u>kg</u> whereas osmolarity is the number of osmols per <u>mole</u>. Both are measures of osmotically active solutes present in serum or urine. Various solutes can contribute to osmolarity including sodium ions, glucose and urea. [☞ Osmolar Gap].

Osmolar Gap

Not to be confused with the anion gap [☞ Anion Gap]. The osmolar gap is the difference between the actual, <u>measured</u> osmolality of urine or serum, and the <u>calculated</u> osmolarity. Osmolarity may be calculated (approximate) from using S.I. units:

$$2 \times [Sodium] + [Glucose] + [Urea] = Osmolarity_{calculated}$$

<u>OR</u> using 'standard' U.S. units:

$$2 \times [Sodium] + \frac{[Glucose]}{18} + \frac{[Urea]}{2.8} = Osmolality_{calculated}$$

Substituting the upper limit reference range values into the above equation yields a calculated osmolarity of around 295 mosmol/L. A large urine osmolar gap may be due to the presence of exogenous substances that contribute to osmolarity, e.g., methanol, polyethylene glycol, etc.

'Osmosis' is derived from the Greek word, 'osmos' meaning, 'a push'. This piece of Chemspeak was invented in 1854 by an Englishman called Thomas Graham (1805-69).

Osmolarity

Not to be confused with osmolality [☞ Osmolality].

PABA Test

Reference range : normal pancreatic function: Index* > 0.76

An *in*direct test of pancreatic function. Specifically a test of the presence of functional chymotrypsin (secreted by the pancreas). The test is very similar to the fluorescein dilaurate test [☞ Fluorescein dilaurate test]. Following an overnight fast the patient is given an oral dose of synthetic polypeptide called BT-PABA. The dose includes 500 mL of water and casein. In a normal healthy person the BT-PABA is cleaved by the enzyme chymotrypsin (secreted by the pancreas) to give BT and PABA. The PABA is absorbed and reaches the liver where it is conjugated. It then enters the general circulation and is excreted in the urine via the kidneys. The excreted PABA can then be measured in the urine. If a reduced amount of PABA is detected then one of the following conditions must be true:

(1) the patient lacks chymotrypsin (or the chymotrypsin is not functioning) due to pancreatic disease etc;
(2) the patient is unable to absorb PABA and transport it to the liver; or
(3) the patient is unable to excrete PABA due to renal disease or urinary tract obstruction.

You can rule out possibilities (2) and (3) by running a 'control' at the same time that you run the test. The 'control' is very simple: the patient is given *radioactively labelled* PABA , i.e., ^{14}C-PABA at the same time that they are given the *non*-radioactively labelled BT-PABA. If *radioactive* ^{14}C-PABA appears in the urine we know that the patient *is* able to absorb ^{14}C-PABA *and* transport it to the liver; *and* that the patient *is* able to excrete ^{14}C-PABA into the urine. If little or no *non*-radiolabelled PABA is found in the urine it means that the patient lacks chymotrypsin (or the chymotrypsin is not functioning) due to pancreatic disease or insufficiency, i.e., they are unable to cleave BT-PABA into BT and PABA. Results of the test are expressed as the PABA/^{14}C-PABA Index*.

$$\frac{\underline{PABA \ excreted}}{PABA \ ingestion} \bigg/ \frac{^{14}\underline{C\text{-}PABA \ excreted}}{^{14}C\text{-}PABA \ ingested} \times 100 = PABA/^{14}C\text{-}PABA \ Index$$

Parathyroid Hormone (PTH)

Reference range: < 120 pg/mL (varies with method/institution)

PTH is secreted by the parathyroid glands which are embedded in, but quite separate from, the thyroid gland [☛ Thyroxine, Triiodothyronine]. Increased PTH may be primary (increased activity stemming from the parathyroids themselves or secondary, due to a decrease in calcium (hypocalcemia) [☛ Calcium]. PTH acts on bone to increase bone resorption (increases serum calcium and phosphate); on kidney (a) stimulates 1α-hydroxylation of vitamin D_3 (indirectly increasing uptake of calcium and phosphate in the gut) (b) increases calcium reabsorption in tubules but decreases reabsorption of phosphate [☛ Phosphate] in tubule (c) decreases bicarbonate [☛ Bicarbonate] reabsorption in tubules which may result in acidosis [Acid-Base]. The effect on tubular reabsorption of calcium is small such that the bottom line is: PTH increases serum calcium phosphate increases renal calcium excretion.

P.C. (Post-Cibum)

After a meal. Not to be confused with Macs (the white plastic ones with 'mice' on mats—not the black shiny ones with rubber boots and hats; nor with the lower ranks of the British police force; nor with political correctness [☛ Humor]

pCO$_2$

Reference range: 4.66- 5.99 kPa {35-45 mmHg}

The partial pressure of carbon dioxide. Partial pressure is the gas world equivalent of concentration [Acid-Base, Blood Gases, pH].

Pentagastrin test

Reference range:
resting juice volume < 50 mL
base line acid secretion: < 5 mmol/hour
post-pentagastrin acid secretion: < 45 mmol/hour (male) < 35 mmol/hour
(female)

A *direct* test of gastric acid secretion. In normal healthy persons the G-cells of the stomach secrete the hormone gastrin into the *blood* stream in response to the presence of proteins, food, etc. in the lumen of the stomach or following stimulation by the vagus nerve. Gastrin travels through the general circulation and stimulates the parietal cells of the stomach to secrete gastric acid (mainly hydrochloric acid (HCL)) and the enzyme pepsin into the lumen of the stomach.

Following an overnight fast the patient has nasogastric tube inserted into the stomach. Base line juices are collected for 1 hour. Pentagastrin, a synthetic analogue of the hormone gastrin, is given intra-muscularly. The contents of the stomach are aspirated for 1 hour. Baseline and post-stimulation samples are analyzed for total volume and acid content (pH) .

High levels of *base line* acid (before giving pentagastrin) are sugges-tive of the Zollinger-Ellison syndrome [☞ Zollinger-Ellison syndrome]. A large volume of *base line* contents suggests gastric stasis.

If the pH > 7 units, following stimulation by pentagastrin, then the patient may be suffering from achlorhydria (absence of HCl in gastric secretions).

'Gastric' is derived from the Greek word, 'gaster', meaning, 'belly'.

pH

Reference range: 35-45 nmol/L {7.35-7.45 units}

$$pH = -\log [H^+]$$

...which is unfortunate. The pH system is counterintuitive at best, and confusing at worst. The <u>higher</u> the concentration of H^+ ions the <u>lower</u> the pH and vice-versa.

Many European hospitals and labs use the concentration of H^+ ions rather than pH because it is NOT intuitively obvious that a change of 0.3 pH units is actually a <u>doubling</u> or <u>halving</u> of H^+ concentration!

An <u>acid</u> is a proton (H^+) <u>donor</u>. A <u>base</u> is a proton <u>acceptor</u>.

The word 'acid' is from the Latin word, 'acidus' meaning 'tart'.

Phosphate

Reference range: 0.8-1.5 mmol/L {2.5-4.6 mg/dL}

PTH [☞ Parathyroid Hormone] has a <u>net</u> effect of decreasing serum phosphate; Vitamin D_3 increases serum phosphate (increased uptake from gut). Hyperphosphatemia may be caused by: renal failure, hypoparathyroidism [☞ Parathyroid Hormone], acromegaly [☞ Growth Hormone], excess intake/administration (iatrogenic), excess vitamin D etc. Hypophosphatemia may be caused by: vitamin D deficiency, primary hyperparathyroidism, malnutrition, inadequate i.v. feeding etc.

Phosphorus

[☞ Phosphate]

Plagiarism

Reference range: whatever can be got away with.

Borrowing other peoples ideas, etc.– and getting caught. Not to be confused with convergent thinking or pure coincidence. For example Newton and Leibnitz simultaneously invented calculus; Darwin and Wallace independently came up with the theory of natural selection [☞ Darlington] or even the striking similarity of a Robert Frost (whoever he was) poem, to one penned by yours truly: Note that Frost claimed never to have visited Darlington, indeed he never mentions Darlington in any of his poems.

Borrowing my Words on a Slow Evening	Stopping by Woods on a Snowy Evening
Whose words these are I think you know His name escapes me even so He will not mind me borrowing here And watch his words fill my quarto	Whose woods these are I think I know His house is in the village though He will not mind me stopping here To watch his woods fill up with snow.
The reader must think it queer To see plagiarism quite so crystal clear I glean his words and perpetrate The starkest theft of my career	My little horse must think it queer To stop without a farmhouse near Between the woods and frozen lake The darkest evening of the year.
I check my words for safety's sake To be sure there *is* some mistake Lest I be sued for sounding the same As you know who - old what's his name?	He gives his harness bells a shake To ask if there is some mistake The only other sound's the sweep Of easy wind and downy flake
His words are subtle, skilled and deep But I have premises to upkeep And much to write before I eat And much to write before I eat	The woods are lovely, dark and deep But I have promises to keep And miles to go before I sleep And miles to go before I sleep.
Ian Wilkinson ©1987	*Robert Frost*

Poetry

Reference range: From every line a rhyme, to random frenetic phonemes produced by an icon*oclastic computer that hates Windows*™ .

An incurable affliction that affects even the most mentally robust people.[☛ Plagiarism, Limerick, Iatrish, Synonyms].

Potassium

Serum reference range: 3.5-5.0 mmol/L {3.5-5.0 mEq/L}

Major intra-cellular ion. Secreted in the distal tubules of the kidney (controlled by aldosterone) in exchange for sodium (retained). Potassium is essential for neuromuscular function. High and low values can lead to cardiac problems, etc.

pO$_2$

Reference range: 10.64- 13.33 kPa {80-100 mmHg.}

The partial pressure of oxygen in the blood. Partial pressure is the gas world equivalent of concentration [☛ Acid-Base, Blood Gases, pH].

'Oxygen' is another piece of Chemspeak. It is mostly that gargantuan Gallic chemist, Antoine Lavoisier's (1743-94) fault (or at least his head's fault: it and he later parted company courtesy of Mdme. Guillotine. 'Oxygen' is a compound of two words: 'oxys', Greek for, 'sharp' and 'gennao' meaning 'I produce' [☛ Carcinoembryonic Antigen] This all sounds very logical and terribly sensible except that it turned out that old Antoine got it wrong: it is hydrogen [Acid-Base] that is the essential constituent of acids, not oxygen.

Predictive Value

A measure of how many truly positive tests (TP) out of the total number of positive tests = true positive (FP) + false positive tests:

$$\text{Predictive Value} = \frac{TP}{TP + FP}$$

Progesterone

Reference range (male, adult): 0.4-1.0 nmol/L {0.13-0.31 ng/mL}
Reference range (female, follicular): 0.06-2.86 nmol/L {0.02-0.9 ng/mL}
Reference range (female, luteal): 19.1-95.4 nmol/L {6.0-30.0 ng/mL}

Serum progesterone assays determine whether or not a woman has ovulated, i.e., is she fertile? Progesterone is produced by the corpus luteum [☛ Luteinizing Hormone]. If the woman is infertile because of lack of ovulation then serum progesterone levels will not rise after day 15 of the menstrual cycle. The corpus luteum is the remains of the follicle after ovulation. Thus, in a sense the corpus luteum is a 'new', temporary 'gland' that appears every month or so during the fertile part of a woman's life. Ovulation is facilitated by the mid-cycle (about day 15) surge in LH and FSH [Follicle Stimulating Hormone]. As its name implies, progesterone prepares for gestation; it prepares the endometrium (lining of the womb) for implantation by the fertilized ovum and prepares the breasts for lactation.

Prolactin

Reference range (male or female): 0-20 μg/L {0-20 ng/mL}

     Prolactin is produced by the anterior pituitary. It facilitates lactation. Hyperprolactinemia is observed in patients with prolactinomas and is common in other pituitary disorders (this is probably because prolactin control is in a negative sense: a Prolactin <u>Inhibiting</u> Factor {dopamine} must reach the pituitary from the hypothalamus in order to inhibit prolactin release. There does not appear to be a prolactin <u>stimulating</u> factor.). Hyperprolactinemia may result in galactorrhea (excessive flow of breast milk). Excess levels of prolactin also inhibit ovulation causing infertility [☛ Progesterone].

     'Lac' is the Latin word for 'milk'. Hence, 'pro-lactin', promoting milk production. A pretty good piece of Chemspeak. A diamond in the rough.

Protein (total)

Reference range (serum): 60-80 g/L {6.0-8.0 g/dL}
Reference range (urine): < 0.15 g/day {<15 mg/dL}

     Serum total protein is of little value alone. One should assess serum albumin [☛ Albumin] and globulins and if appropriate run protein electrophoresis.

     Urine protein should be < 0.15 g/day [☛ Urine collection, 24 hour]. Dipsticks used routinely screen urine samples will show negative for urinary protein unless protein is above reference range. Increased protein may be observed in: nephrotic syndrome, glomerulonephritis, pyelonephritis, multiple myeloma benign postural proteinuria etc.

     'Protein' is pure Chemspeak. It was invented in 1838 by that dapper daguerrogenic Dutch chemist Gerard Mulder (1802-80). He believed that proteins were the essence of living matter. The word was stolen and adapted from the Greeks: 'proteios' meaning 'first place'.

PTH

[☛ **Parathyroid Hormone**]

Renin

[☞ Renin-Angiotensin System]

Renin-Angiotensin System

A fall in blood volume/pressure is detected by the juxta-glomerular apparatus of the kidneys, causing release of renin. Renin converts angiotensino**gen** to angiotensin which is further metabolized in the lungs. Angiotensin stimulates the production of aldosterone [☞ Aldosterone] which stimulates retention of sodium in the distal tubule. Sodium retention leads to increased water retention and restoration of homeostasis. In addition, anti-diuretic hormone [☞ Anti-Diuretic Hormone] is released and the thirst centers stimulated.

Reference Range

A range of values within which the majority of healthy patient's values lie. In a perfect world this range would be symmetrically distributed about the mean—the familiar 'bell-shaped' curve. A completely separate curve of patients who *do* have the disease would have a mean higher (or lower) than that of the reference range. There would be no overlap between the two curves. Alas! Alack! In the real world, the 'healthy patients' population curve and the 'sick' patients' population curve, are often asymmetrical and do overlap. [☞ Sensitivity, Specificity, Predictive Value, Efficiency].

Reference ranges can be influenced by many factors: age, gender, race, diurnal and seasonal rhythms, menstrual cycle etc.

The term 'reference range' is preferred over the older term: 'normal value or range'. What is normal? An average person? (Everyone wants to be 'normal' but no-one wants to be 'average'—though half of us must be below average by definition). A line drawn perpendicular to another? A one molar solution? Someone who is mentally robust?

Reverse T$_3$

[☞ Reverse Triiodothyronine]

Reverse Triiodothyronine (Reverse T$_3$)

Reference range: 0.46-1.23 nmol/L (30-80 ng/dL)

Physiologically inert analogue of T$_3$ [☞ Triiodothyronine]. Both T$_3$ and Reverse T$_3$ are derived from T$_4$ [☞ Thyroxine]. The amount of each type of T$_3$ produced seems to be affected by: (increased Reverse T$_3$): acute stress, starvation etc. Increased Reverse T$_3$ plus low T3 is suggestive of sick euthyroid syndrome (the patients' abnormal thyroid hormones are not due to a primary thyroid disorder).

Secretin-CCK Test

[☞ Cholecystokinin-secretin Test]

Sensitivity

A measure of the number of truly positive tests (TP) out of all those tested who do have the disease = TP + false negative (FN):

sensitivity = $\dfrac{TP}{TP + FN}$

SIADH

[☞ Syndrome of Inappropriate secretion of Anti-Diuretic Hormone].

Signs

Not to be confused with symptoms [☞ Symptoms]. Signs are objective, observable manifestations of disordered functioning of the body. For example the *signs* of dehydration may include: dryness of the mucous membranes, decreased salivary secretions, loss of skin turgor, decreased urine volume and weight loss.

Sodium

Serum reference range: 135-145 mmol/L {135-145 mEq/L}
Urine reference range: 40-210 mmol/day {40-210 mEq/day} (varies with diet)

The major extracellular ion present in ECF. Reabsorbed in tubules. "Fine tuning" in distal tubule controlled by aldosterone (sodium reabsorbed in exchange for either potassium or hydrogen ions). Water follows sodium. To increase ECF volume the body retains sodium. To reduce ECF volume the body excretes sodium. Sodium. potassium, bicarbonate and chloride are the so-called electrolytes.

Somatotropin

[☛ Growth Hormone (GH)]

Specificity

A measure of the number of truly negative (TN) out of all those tested who do NOT have the disease = TN + false positive (FP):

$$\text{specificity} = \frac{TN}{TN + FP}$$

Symptoms

Not to be confused with signs [☛ Signs]. Symptoms are the manifestations of disordered functioning of the body. Often these are apparent to the patient and may be subjective. For example, the *symptoms* of dehydration may include: thirst, a feeling of a dry mouth; difficulty in swallowing, weakness and confusion.

Syndrome

A group of symptoms and signs that are frequently associated with one another. The term 'syndrome' tends to be used when the full picture is not understood or clear. Syndromes are often named after obscure people whose names take up more of your memory, give no information about the condition and are often difficult to spell: e.g., Crigler-Najaar, Dubin-Johnson and Zollinger-Ellison syndromes.

Wilkinson's Syndrome, also known as, Eponymous Nuncupatic Antonomasia afflicts 90% of people who do not have any syndrome named after them.

The Greek words, 'syn' means, 'in connection with' and 'dromos' means, 'a course'. Thus, a syndrome is a group of symptoms and signs that 'run a course together'.

Syndrome of Inappropriate Secretion of Anti-Diuretic Hormone (SIADH)

Release of ADH [☞ Anti-diuretic Hormone] in inappropriate circumstances; due to: (1) Ectopic secretion: bronchial carcinoma, cancers of the thyroid and prostate etc releasing ADH-like factors. (2) inappropriate secretion: pulmonary causes (pneumonia, TN, ventilation); cerebral causes (head trauma, tumors, aneurysms etc); other causes (pain, Guillain-Barre syndrome, hypothyroidism, narcotics). The exact means by which pituitary ADH secretion is stimulated is not fully understood. Edema is not seen in SIADH, since the increased water is distributed between the vascular and interstitial compartments.

Synonyms

Not to be confused with a type of sticky bun. A cinnamon bun is just that, whereas a synonym bun is also a cinnamon bun but with a different name, that means more or less the same thing. Pharmacists and pharmaceutical companies like to give drugs lots of different names; partly to be consistent with the inconsistencies consistently observed in Iatrish and Chemspeak and partly for fun. [☞ Drugs].

If still unclear consider the following:

Honorificabilitudinatibus

Its a new, new, new, new langue du pays
(saves on reallylongwords and hyperbole)
Needs no hyperlatives (the literatis' psy-purgatives
Nor classical allusions of Agonippean thrust
Nor foreign phrasing replacing some English mot juste
Dictionaries overawe me., and 'ditto' Thesauri
(Just refuse to confuse the incogniescenti)
So misuse and abuse them and throw 'em away.
Its no fouler than Fowler to treat words this way
"But does this not make one's language a bore?"
Asked the marble-mouthed Marquis on 'The Culture-Vulture Show'
(Though not really in those words: his weren't onamatopeic,
but snotty, and dead clever and archaic!
(I could say that word - if this were art
(but words are the Molotovs of the smart).
"An ongoing discontinuance of sesquipedalianism" I sing
"For the meaning of anything is just another word for the same thing."

Ian Wilkinson © 1982, 1990

T₃

[☛ Triiodothyronine]

T₃ Uptake

[☛ Triiodothyronine Uptake]

T₄

[☛ Thyroxine]

Tamm-Horsfall Glycoprotein

A normal (non-pathological) constituent of urine. Urine collected first thing in the morning often contains this glycoprotein. It is derived from cells in the lumen of the kidney (mainly in the Loop of Henle). The sheath of casts are made from Tamm-Horsfall glycoprotein. [☛ Casts, Urinalysis].

TBG

[☛ Thyroxine Binding Globulin]

Testosterone (total)

Reference range (male): 9-30 nmol/L {260-865 ng/dL}
Reference range (female): 0.5-2.5 nmol/L {14-72 ng/dL}

This test is useful for investigation of hypogonadism in males and for investigation of hirsutism/ and/or virilization in females (suggests cause is ovarian rather than adrenal). Hormone levels vary widely. Three equally spaced samples should be obtained and pooled. Testosterone binds to a sex hormone binding globulin (SHBG) in serum. It is the free (unbound) testosterone that is physiologically active. Certain conditions, such as hyperthyroidism may increase the amount of (SHBG) in serum and therefore reduce the amount of free testosterone resulting in hypogonadism.

'Testosterone' is 'a steroid from the testis'. 'Testis' is the Latin word for 'witness'. So what is the connection? Apparently, in ancient times, before the advent of indecency laws and the Thought Police, men would, publicly, grasp their scrotum (and on occasion other men's scrotums {scrota? scrotae?}) and swear a 'testi-mony'. It all started very early on, in the Bible. "And the servant put his hand under the thigh of Abraham his master and swear to him concerning that matter." Genesis 24:9.

TG

[☛ Triglycerides]

Thyroid Stimulating Hormone (TSH)

Reference range: < 5 mIU/L {<5 µIU/mL

TSH is released by the anterior pituitary and stimulates the thyroid to synthesize and release T_4 [☛ Thyroxine]. TSH is screened for, in neonates to detect and treat, potential cretinism (congenital thyroid deficiency resulting in arrested mental and physical development; and bone dystrophy). TSH is the most sensitive test for primary hypothyroidism (TSH = elevated; in secondary hypothyroidism {deficiency of TSH} and in hyperthyroidism {excess T_4 causes feedback suppression of TSH} TSH is low).

Thyroxine (T_4)

Reference range (total): 60-150 nmol/L {4.7-11.6 µg/dL}
Reference range (free): 9-26 pmol/L {0.7-2.0 ng/dL}

Thyroxine is released from the thyroid in response to stimulation by TSH [☛ Thyroid Stimulating Hormone]. Thyroxine are essential for normal growth and development, having a wide range of effects upon various metabolic processes. Congenital deficiency of thyroid function results in cretinism (neonates are screened [☞ Thyroid Stimulating Hormone]). Thyroxine (T_4) is converted to T_3 [☞ Triiodothyronine] systemically. T_3 is believed to be the physiologically important entity. In certain conditions T_4 is converted to the physiologically inert reverse T_3 [☞ Reverse Triiodothyronine].

'Thyroxine' and 'thyroid' are derived from the Greek word 'thyreos' which originally meant, 'a large stone used to keep a door shut' and later it was extended to refer to a large oblong shaped shield, that had a notch a the top. So what is the connection? Apparently these shields resemble the notched laryngeal cartilage. The thyroid gland itself looks more like a butterfly than a shield. I guess people just had more imagination in those days—look at what they saw in the constellations: bears, hunters and twins; all I can see when I look at them are polygons, tax forms, and the occasional supernova explosion.

Thyroxine Binding Globulin (TBG)

Reference range: 15-34 mg/L {1.5-3.4 mg/dL}

As its name implies TBG binds T_4 [☞ Thyroxine] and T_3 [☞ Triiodothyronine]. More than 99% of these hormone is protein bound (to TBG, albumin and prealbumin {also called 'transthyretin'}). Only unbound (free) hormone is physiologically important. Thus, any change in the amount of TBG (and prealbumin, albumin) will increase or decreases the amount of free T_4 and free T_3. TBG may be increased (free hormones therefore decrease = potentially hypothyroid) by: contraceptives, estrogens, pregnancy (increased synthesis), congenital excess TBG. TBG may be decreased (free hormone levels increase = potentially hyperthyroid) by: major illness, surgery, malnutrition, malabsorption, nephrotic syndrome, congenital conditions etc.

Total Iron Binding Capacity (TIBC)

Reference range: 45-80 µmol/L {251-447 µg/dL}

Measurement of serum iron alone is of limited use as a means of assessing iron status. Iron levels fluctuate widely during and between days. Iron is stored as ferritin [☞ Ferritin] and transported by transferrin. Each transferrin molecule has two iron binding sites. In health, on average 33% of these sites are occupied. It is now possible to directly measure transferrin concentration. TIBC is an indirect estimate of transferrin levels. Serum iron is firstly measured. Then excess iron is added to the sample to fully saturate any unoccupied transferrin iron binding sites. Unbound excess iron is then removed (e.g. using a chelating agent such as magnesium carbonate) and iron levels are once again determined. The original percent binding of iron can then be calculated. Hematological tests are an essential part of any full assessment of iron status.

Triglycerides (TG)

Reference range: < 1.8 mmol/L { <159 mg/dL} (varies with age, gender, diet etc)

TGs (fats) are insoluble in water and are, therefore, transported, in the blood associated with protein/cholesterol/TG (chylomicrons and Very Low Density Lipoproteins (VLDL). Following a meal it takes approximately 12 hours to clear the blood of chylomicrons, thus a fasting (> 12 hours) is required. Increased TGs may be observed in: primary hyperlipidemia, diabetes mellitus [☛ Diabetes Mellitus], acute alcoholism, oral contraceptives, nephrotic syndrome, chronic renal failure, steroids, glycogen storage disease.

Triiodothyronine (T$_3$) (total)

Reference range: 1.2-2.9 nmol/L {78-188 ng/dL}

T$_3$ is the physiologically active thyroid hormone. It is synthesized directly in the thyroid and also produced systemically by the conversion of T$_4$ [☛ Thyroxine]. T$_3$ binds to TBG [☛ Thyroxine Binding Globulin]. Only unbound (free) T$_3$ is physiologically active.

Triiodothyronine (Resin) Uptake (T$_3$ Uptake)

Reference range: 0.24-0.34 fraction {24-34}%} of total

This test is an estimate of how many unoccupied binding sites are present on thyroxine binding proteins [☛ Thyroxine Binding Globulin]. Because only unbound (free) thyroid hormones are physiologically active [☛ Thyroxine, Triiodothyronine], measurement of T$_3$ Uptake is useful in assessing thyroid function.

Decreased uptake may be observed because TBG etc are decreased (androgen excess, renal failure, malnutrition, congenital low TBG etc) or because T_3 binding sites are occupied by something else (T_4, dilantin, salicylate etc). Increased uptake is observed in the opposite situations: estrogens, pregnancy etc.

Triolein breath test

Reference range: determined by each lab

An *in*direct test of fat absorption. Normal healthy persons break down fats in the gut via the action of enzymes such as lipase (secreted from the pancreas). Bile salts are essential for micelle formation. Micelles are spherical groupings of ingested fats, bile salts and enzymes such as lipase. Patients suffering from steatorrhea (> 7 grams of fat per day excreted in the feces [☞ Fecal fat]) are unable to absorb fats efficiently. The triolein breath test uses a radioactively labelled synthetic fat to measure the degree of fat absorption.

Following an overnight fast a base line breath sample is collected from the patient. The patient is then given an oral dose of radioactively labelled triolein (^{14}C-triolein) in a 60 gram fat meal. Expired air is then collected hourly for 7 hours. In normal healthy patients the ^{14}C-triolein will eventually arrive in the small intestine where it will be broken down into glycine and free fatty acids by the action of lipase. The glycine and free fatty acids are eventually absorbed by the body and metabolized to $^{14}CO_2$ and other products. The radioactively labelled $^{14}CO_2$ is eventually excreted in expired breath.

In patients suffering from steatorrhea, the ^{14}C-triolein *cannot* be broken down and/or absorbed in the lumen of the gut. The ^{14}C-triolein continues down the gastrointestinal tract and is excreted in the feces. Note that in most patients suffering from steatorrhea there is *some* break down of ^{14}C-triolein to $^{14}CO_2$ which is detected in expired breath.

TSH

[☞ Thyroid Stimulating Hormone]

Tube Top Color Guide

Specimens of blood are collected into vacuum tubes which have color coded rubber stoppers. Each color indicates the type of preservative (if any) contained within. Color codes may vary:

GREEN contains heparin. Heparin prevents clotting. Used for tests requiring plasma.

GREY contains fluoride. Fluoride inhibits glycolysis. Used for glucose analysis.

LAVENDER contains EDTA. Prevents clotting. Used mainly by hematology for CBC's.

RED contains no preservatives. Used for tests requiring serum.

Units

Do you still confuse your kilograms and anagrams? Do you forget exactly how many nanometres there are in a speedometer? Ditto kilojoules and killer whales. From 'teras' to 'femtos', here is what's what and why:

Math	English	Prefix	Origin
10^{12}	trillion	tera	Greek. teras: 'monster'
10^9	billion	giga	Greek. gigas: 'giant'
10^6	million	mega	Greek. megas: 'large' P
10^3	thousand	kilo	Greek. chilioi: '1,000'
10^2	hundred	hecto	Greek. hekaton: '100'
10^1	ten	deca	Greek. deka: '10'
10^{-1}	tenth	deci	Latin. decimus: '1/10th'
10^{-2}	hundredth	centi	Latin. centum: '1/100th'
10^{-3}	thousandth	milli	Latin. millesimus:'1/1000th'
10^{-6}	millionth	micro	Italian. pico: 'small'
10	billionth	nano	Greek. nanos: 'dwarf'
10	trillionth	pico	Italian. pico: 'small'
10	quadrillionth	femto	Danish. femten: '15'

Urate

[☞ Uric acid]

Urea

Reference range: 1.3-3.0 mmol/L {8-21 mg/dL}

Urea is useful in checking on kidney function [☞ Creatinine]. Urea is produced in the liver from proteins and amino acids. Increased urea may be observed in renal disease (especially pre-renal), high protein diet etc. In general there are three causes of increased urea: (1) increase intake of urea (protein, amino acids, cell breakdown). (2) decreased glomerular filtration rate (GFR) (hemorrhage, shock, diarrhea etc). [☞ Glomerular Filtration Rate]. (3) Increased reabsorption of urea in the kidney tubules (dehydration).

Urea may be decreased in liver disease (especially chronic), normal pregnancy etc.

'Urea' comes from the French word, 'uree', a name for the major salt found in urine. It is in turn related to the word, 'urine' (from the Greek word, 'ouron', meaning 'urine'. Uremia is 'Iatrish' for 'increased toxic nitrogenous compounds present in the blood'.

Uric acid

Reference range (male) : 0.27-0.48 mmol/L {4.5-8.2 mg/dL}
Reference range (female): 0.18-0.38 mmol/L {3.0-6.5 mg/dL}

Uric acid is derived from the breakdown of purine nucleotides which are in turn derived from (1) synthesis, (2) breakdown of tissues and (3) the diet. Note that the nucleic acids adenine and guanine are both *purines*. Do not confuse with the nucleic acids thymine, cytosine and uracil which are all *pyrimidines*. Uric acid is excreted via the kidneys (about 66%) and via the gut (about 33%). Within the gut bacteria breakdown uric acid (uricolysis) to form CO_2 and NH_3. Increased uric acid is associated with gout. An increase may also result from increased tissue breakdown (leukemia, anemia, infection, chemotherapy, other causes of necrosis etc), chronic renal disease, certain drugs, malnutrition etc.

Urinalysis

The analysis of urine has a long and bizarre history. Modern techniques can be divided into two broad areas: chemical analyses and microscopic analyses. Chemical analysis in the first instance is generally done by means of dipsticks: multi-reagent strips that are dipped into aliquots of urine. These strips are designed such that color changes indicate approximate amounts of substance detected in the urine. Common tests available include: glucose, protein, blood, leukocytes, specific gravity, urobilinogen, ketones, hemoglobin, pH and bilirubin (conjugated).

Microscopic analysis should be performed on a fresh (< 2 hours old) specimen. Elements of interest include: casts, red blood cells (erythrocytes), white blood cells (leukocytes), bacteria and crystals. Many laboratories no longer offer microscopic analyses of urine unless specially requested due to the time and labour involved, the difficulty of interpretation and the power of simple dipstick tests.

Historically urinalysis or rather uroscopy occupied much of a physician's time. Many physicians believed that it was not necessary to actual see or examine the patient—just their urine. Bubbles, sediments, color, smell and taste were all thought essential parts of the analysis. The golden age of urine tasting, etc. is now over—at least as a medically valid procedure.

Urine collection, 24 hour

First day:
> 8:00 patient must empty bladder completely into toilet. This urine is NOT collected.

> collect ALL subsequent urine passed including...

> Next Morning:
> 8:00 collect urine.

Urine collection, random sample

Discard the first 15-20 mls of urine. Collect approximately midstream flow. Do not collect the last few mls of urine.

So much for theory. Patients have been known to do bizarre things with their urine collection: adding colored water, someone else's urine, adding their dog/cat/aardvaak's urine; urinating into the toilet bowl and then somehow (syphon?) filling the collection bottle with the resulting mix of urine, water, bleach, blue dye, toilet paper etc. In addition certain specialized tests require that the urine be kept acidified whilst being collected. For these types of tests, collection vessels contain a small amount of concentrated acid. Despite clear instructions, patients may discard the acid and then start (uselessly) collecting their urine.

Vasopressin

[☞ Anti-diuretic Hormone]

Wilkinson

Reference range: slightly strange to el bizzaro with at least 3 deviants from the mean

Fundamental Darlingtononian unit of Somethingorother. There are several sub-types (derived from the fusion of parent particles known as the 'Lorna' and the 'Archie') including Ann, David Ian, Jane and Peter (the latter two are twin-particles), ranked in order of decay. The sub-type 'Ian' is often confused with large expanses of time (e.g. "life began ians and ians ago..." etc.) and/or with very small charged particles (e.g. anians and catiians; the ianosphere; ianing one's shirt...etc.). The composition of sub-type 'Ian' varies over time–*usually* being made up of any three of the 6 quarks: 'Up', 'Down', 'Truth' 'Strangeness', 'Beauty' and 'Charm'—the uncertainty being due to the rumour that Heisenberg Probably Rules O.K. The only other known facts about this particle are:

Carbon-dating reveals an age of approximately thirty-something . My wife and I are both married, have no children that we know of and hate flossing our teeth. I was born in Darlington, England and arrived in Canada in 1982, having taken the wrong bus in Piccadily Circus. I took a B.Sc and M.Sc. in England, a Ph.D in Manitoba, a D.Clin.Chem. in Toronto; the Canadian National Certification Examinations (FCACB); part 1 of the Royal College examinations (MRCPath.) and am about to take the U.S. Boards examinations—and am presently taking an MBA (part-time) at the U. of T..and a Dip. Clin Admin. (OHA) at work. My day-job is as a Clinical Chemist at Sunnybrook Health Sciences Centre. My favourite 'extra-curricular' interests include, teaching, writing, opera (pronounced 'Uproar'), making beer & wine, riding horses and motorcycles, SCUBA diving, wearing black & not getting shaved on weekends, Chrissie Hinde, Betty Page and coffee. Ambition: to invent a mnemonic for spelling the word mnemonic, be able to sign my name using 'joined-up' writing; to plant a tree, write a book (done! ☞) and have one or three F_1 hybrids.
[☞ Clinical Chemist, Cynical Chemist, Darlington, Jargon]

Xylose Test

Reference range: urine xylose: > 8 mmol/5 hrs {1.2 g/5 hrs.} serum xylose: > 1.3 mmol/L {> 20 mg/dL} at 1 hour after beginning the test.

People do not make xylose; but plants do. Nor do people metabolise xylose. If all is well, healthwise, xylose is absorbed by the cells of the small intestine and it enters the hepatic portal vein and arrives in the liver. It then enters the general circulation and is eventually excreted via the kidneys and appears in the urine. If the patient is suffering from malabsorption (caused, for example, by Celiac disease; Non-tropical sprue; Gluten sensitive enteropathy; etc) then little or no xylose will appear in the urine. Instead the xylose will be excreted in the feces.

Following overnight fast the patient must empty their bladder. The patient is given 5 g of xylose in water and urine collected for 5 hours. The patient must drink at least 500 mL of water within 2 hours of taking the xylose. Note that if *urinary* xylose is measured then you are making the assumption that liver and kidney function are both normal. If, however, you measure *serum* xylose then you are assuming only that liver function is normal.

Xylose is derived from the Greek, 'xylon', meaning, 'wood', and the '-ose' bit is Sciencespeak for 'sugar' Hence, xylsose is 'wood sugar'.

Zollinger-Ellison Syndrome

Hypergastrinemia (high levels of the hormone gastrin in the blood) caused by gastrin secreting tumours. This leads to recurrent duodenal ulcers. Inheritance is autosomal dominant. The tumours are usually found in the walls of the *duodenum*, not in the *stomach* walls. Remember that in normal healthy people gastrin is secreted by the G-cells of the gastric (stomach) epithelium. [☞ Pentagastrin Test].

R.M. Zollinger was born in 1903 in Ohio. Served in the army medical corps (Colonel) and published more than 340 articles. E.H. Ellison (1918-1970) was also born in Ohio. He wrote more than 130 articles. Mr. Spock was born on Vulcan. Commander Riker wasn't. He was born in Alaska. Neither of them published much. Meanwhile, 'Live long and Be Preposterous!'

Acknowledgments

The following texts were drawn upon in writing this book:

'Medical Meanings', William S. Haubrich, Harcourt Brace Jovanovich, New York, 1984

'The ABC's of Interpretive Laboratory Data', Second Edition, Seymour Bakerman, 1984.

'Clinical Guide to Laboratory Tests', Second Edition, Norbert W. Tietz, WB Saunders, 1990.

'Illustrated Textbook of Clinical Chemistry', William J. Marshall, Lippincott, 1989.

'Methods in Clinical Chemistry', A. J. Pesce and L. A. Kaplan, Mosby, 1987.

Suggestions, Improvements, Comments

If you have found this guide to be useful I would very much appreciate hearing from you. Any suggestions, etc. that you have will help me expand and improve subsequent editions of this book. You can contact me at:

Dr. I. Wilkinson
Department of Clinical Biochemistry
Sunnybrook Health Science Center
2075 Bayview Avenue
Toronto, Ontario
Canada, M4N 3M5.
(416) 460-4991
FAX: (416) 480-6120

—NOTES—

—NOTES—